Nature's Enigmas: Untangling the History and Meaning of Foliate Carvings

Haddin

Table of Contents

Preface

Once bedecked with sculpture, the portals of Noyon Cathedral were scraped bare during the political upheaval of the French Revolution (fig. 0.1, 0.2). What remains between the ghost outlines of voussoirs and horizontal registers of the tympanum is the foliate carving; while the figurative sculpture was systematically chiseled from the surfaces of the three portals, the iconoclasts left behind winding vine friezes, capitals covered in vine leaves, a grid of foliate diaper relief, and a single vine sprouting below the trumeau figure of the central portal, damaged in German bombardments. The erasure of the figurative sculpture highlights the presence of foliate forms of the portals' decoration. The revolutionaries, going so far to scrape the reliefs flat from Noyon's quatrefoils, kept the sculpted leaves and vines crisp.

In its current state, the foliate carving at Noyon has been decontextualized from its original format. Other monuments that have endured periods of iconoclasm follow a similar pattern: at Wells Cathedral, figurative sculpture on the west façade was damaged heavily in religious and political upheaval while the foliate elements remained relatively untouched. While the damage and erasure of some portal sculpture make its survival at other sites, such as Reims and Amiens, even more extraordinary, foliate forms were not employed in every sculptural ensemble or architectural context. Even though foliate forms are often associated with Gothic architecture, it seems that they were used judiciously with purpose and intention. Specific buildings have what could be called a distinct foliate program, while others do not. Far from a homogenous enterprise, foliate decoration was personalized to every building site in terms of its form, placement, scale, and relationship to local cults of saints or lore.

Nor were these choices dictated purely by stylistic preferences. In the twelfth century, religious communities applied a wide range of architectural decoration to sacred architecture in accordance with the community's values and priorities. The Cistercian abbey church of Notre-Dame in Pontigny, for example, has no sculptural decoration in the nave, foliate or figurative, a choice reflecting the worldview of the Cistercian enterprise of the first half of the twelfth century (fig. 0.3).[1] In contrast, the nave and narthex of La Madeleine at Vézelay, a pilgrimage destination rebuilt at great expense, are full of diverse carved foliate bands, in addition to a frieze of repeating foliate motifs that runs along the exterior (fig. 0.4). At Vézelay, the attention to foliate decoration, created in concert with the figurative capitals and portal sculpture, appears in concert with a diverse audience that extends beyond the local religious community to include laypeople and traveling pilgrims.

In Gothic architecture of the late twelfth and early thirteenth centuries, not all buildings incorporated foliate sculpture. The later chevet at Vézelay, for example, stands in contrast to the richness of the earlier foliate carving; its Gothic chevet built between 1165-1180 does not reference the foliate motifs or rich sculpted decoration present in the rest of the church.[2] In the relatively contemporary Gothic chevet at Pontigny, the crocket capitals of the Gothic chevet introduce elements of foliate sculpture, albeit abstract, into the otherwise austere building.[3] Many

---

[1] Terryl N. Kinder, "Toward Dating Construction of the Abbey Church of Pontigny," *Journal of the British Archaeological Association* 145 (January 1, 1992): 77. See also Jean Owens Schaeffer, "The Earliest Churches of the Cistercian Order," in *The New Monastery: Texts and Studies on the Earliest Cistercians*, ed. E. Rozanne Elder, Cistercian Fathers Series, no. 60 (Kalamazoo, Mich.; Spencer, Mass: Cistercian Publications, 1998), 195–208.

[2] Arnaud Timbert and Philippe Montier, *La cathédrale Notre-Dame de Noyon et son quartier* (Noyon: Cap régions, 2012).

[3] Cynthia Marie Canejo, "Transforming Early Gothic Form: The Cistercian Church of Pontigny, Saint-Martin at Chablis, and Northern Burgundian Architecture" (Ph.D., California, University of California, Santa Barbara, 2005), 28-33. The chevet at Pontigny was likely begun after Vézelay but might have already been underway before 1180.

early iterations of Gothic architecture from this period limit decorative sculpture to crocket capitals, though this is not always the case. At Notre-Dame of Paris, the nave capitals present a wide variety of foliate forms; the same is true for Wells and even Canterbury, a point that will be explored in more detail in this dissertation. Nor is it always true that thirteenth-century Gothic buildings incorporate foliate sculpture with increasing frequency, as many scholars have argued. In England, the architects of the nave of thirteenth-century Salisbury cathedral elected for turned capitals and minimal foliate sculpture on the interior, standing in contrast to the architectural decoration in the contemporary work at Wells and Lincoln. Regionality may account for some of this variation, but the surviving material record of sacred architecture from the twelfth and thirteenth centuries suggests that builders and sculptors also exercised choice and creativity in selecting modes of architectural decoration.

In certain locales, this type of foliate sculpture took on a new monumentality and aesthetic complexity. Foliate elements increased in number, size, length, and formal complexity, occupying an expanded field of decorative possibilities. At Notre-Dame of Amiens, foliate sculpture not only playfully interacts with the architectural frame of the building but enhances the iconographic program of the west façade and resonates with the local patron saint, Firmin the Martyr, whose open tomb turned a winter's day into spring, prompting trees to burst into bloom. At Wells Cathedral, one of the earliest examples of Gothic architecture in England, foliate decoration articulates the quire, nave, and west façade in distinctive ways not repeated in other English cathedrals. For both bishoprics, images of foliage were integrated into the civic identity of the surrounding town: the city of Amiens adopted sprigs of foliage into its coat of arms at an unknown date, but early examples can be found in manuscripts from at least the sixteenth century; and at Wells, an image of a tree growing out of the eponymous wells was integrated into the cathedral's seal in the thirteenth century and adopted by the borough in the civic coat of arms.[4] The presence of foliage in the civic identity and the presence of this type of decorative ornament in these cathedrals raises questions about whether foliage played a more active role in solidifying a unique architectural and civic identity for these buildings. Perhaps foliate forms, if attended to, may reveal more than aesthetic enrichment in certain contexts.

While foliate forms are frequently commented upon in studies of Gothic architecture, they have not received the same critical treatment as other types of sculpture or architectural decoration, even though many examples appear ostentatiously at eye-level or have been designed in such a way to be seen and understood from a great distance. A challenging aspect of studying foliate sculpture from the twelfth and thirteenth centuries is that the elaborate foliate sculpture of the later Middle Ages, such as the sixteenth-century *astwerk* of Ulm Minster or the later example at Liebfrauenmünster Ingolstadt tend to make earlier examples appear tamed or controlled in comparison (fig. 0.5, 0.6). Matt Kavaler acknowledges this comparative gaze by presenting late Gothic iterations of foliate sculpture as disruptive, in which the "willful and organic seem to endanger the precise and predictable" and engage in "a larger struggle between order and chaos, a commentary on the limits of human control."[5] When referring to thirteenth-century examples, Kavaler writes that these forms are sequestered to "discrete areas and fields," playing by an unwritten set of rules, in contrast to their "disruptive" counterparts. Writing about Southern Germany, Paul Crossley acknowledges that foliate sculpture found in late Gothic German churches seems to take inspiration from fourteenth-century French secular architecture, but a longer view of foliate sculpture reveals that the forms adopted in secular architecture actually

---

[4] Warwick Rodwell, *Wells Cathedral: Excavations and Structural Studies*, 1978-93 (London. English Heritage, 2001), 130-33, 375-76.
[5] Ethan Matt Kavaler, *Renaissance Gothic: Architecture and the Arts in Northern Europe, 1470-1540* (New Haven: Yale University Press, 2012), 201

expanded upon ideas from twelfth- and thirteenth-century churches, which were not only

experimental for this medium but consequential.[6] Looking at foliate sculpture in context allows

the experimental qualities of earlier examples to come into focus.

Since vegetal decoration can nimbly refer to a number of Biblical texts, revisiting

medieval commentaries on church decoration provides additional context for understanding the

overlapping of foliate forms and architecture during this period. Bernard of Clairvaux's *Apologia*

of 1125 addresses religious sculpture directly, but also differentiates the role of art in the

monastery from that of the cathedral: "For certainly bishops have one kind of business and

monks another. We know that since they are responsible for both the wise and the foolish, they

stimulate the devotion of a carnal people with material ornaments because they cannot do so with

spiritual ones."[7] Bernard levels his critique at art for the use of monks in the monastery, rather

than the decorations that adorn cathedrals for the purposes of conveying religious ideas to

laypeople.[8] That latter category of decoration, according to Bernard's text, has a clear function of

"stimulating devotion." Cathedral decoration that included vegetal ornament in addition to

column figures, sculpted tympana, trumeau figures, and scenes sculpted in relief could enhance

the experience of the Church as an other-worldly realm—as noted above, the Cistercian abbey at

---

[6] Paul Crossley, "The Return to the Forest: Natural Architecture, the German Past in the Age of Dürer," in *Künstlerischer Austausch: Akten Des XXVIII Internationalen Kongresses Für Kunstgeschichte, Berlin, 15-20 Juli, 1992*, ed. Thomas W. Gaehtgens, vol. 2 (Berlin: Akademie Verlag, 1992), 71–80.

[7] "Scimus namque quod illi, sapientibus et insipientibus debitores cum sint, carnalis populi devotionem, quia spiritualibus non possunt, corporalibus excitant ornamentis." *Library of Latin Texts - Series A*, accessed August 25, 2020. Trans., C. Rudolph, *Bernard*, p. 10.

[8] Art historians' engagement of St. Bernard's *apologia* has a rich historiographic record. See Meyer Schapiro, "On the Aesthetic Attitude in Romanesque Art," in *Romanesque Art* (New York: G. Braziller, 1977), 1–27; Erwin Panofsky, *Abbot Suger on the Abbey Church of St.-Denis and Its Art Treasures* (Princeton, N.J.: Princeton University Press, 1979), 6; Conrad Rudolph, *The "Things of Greater Importance": Bernard of Clairvaux's Apologia and the Medieval Attitude toward Art* (Philadelphia: University of Pennsylvania Press, c1990), 6; Thomas E. A. Dale, "Monsters, Corporeal Deformities, and Phantasms in the Cloister of St-Michel-de-Cuxa," *The Art Bulletin* 83, no. 3 (2001): 402–36; Mary J. Carruthers, "'Varietas': A Word of Many Colors," *Poetica* 41, no. 1/2 (2009): 11–32, esp. 18.

Pontigny, not built for pilgrims or a diocesan seat of power, had very little foliate ornament, in contrast to Vézelay. Stone sculpted to look like other materials, whether vines, leaves, or abstract foliate elements such as crockets, transformed architectural materials into a mutable, organic substance that could attract the gaze and wonderment of viewers. The approach to architectural decoration in sacred space seems to reflect the status of a building with its public, real or imagined. This public might include, or aspire to include, local inhabitants, members of the clergy of surrounding parishes, pilgrims, secular political leaders, as well as abbots or other clergy in roles of leadership in the church.

Abbot Suger, whose texts on the rebuilding, decoration, and consecration of the abbey church of Saint-Denis are often painted as a direct response to Bernard's call for artistic restraint, addresses other aspects of church decoration.[9] Unlike Bernard, he does not address architectural sculpture directly; he mentions capitals briefly, but does not comment on the quality of their carved decoration or any aspects related to their meaning—Gervase of Canterbury, in comparison, specifically calls out the quality of the capital carving in the rebuilt sections of Canterbury Cathedral.[10] Suger is absorbed, rather, in cataloging precious materials used in the church's altars, reliquaries, and mosaics, even as he claims that *materiam superabas opus,* "the work surpassed the material," an Ovidian reference also used by Gervase of Canterbury and

---

[9] Lindy Grant, *Abbot Suger of St.-Denis: Church and State in Early Twelfth-Century France* (New York: Longman, 1998), 4-6. On the rhetorical structure of Suger's writings and its relationship to the rebuilding of Saint-Denis, see Stephen Murray, "Suger, Abbot of S-Denis, and the Rhetoric of Persuasion," in *Plotting Gothic* (Chicago: University of Chicago Press, 2014), 73–95. The lively debates over Abbot Suger's writings largely emanated from Erwin Panofsky's discussion of Suger's theological and theoretical concerns of the aesthetics of Saint-Denis. For an overview of these responses.

[10] For example, Suger mentions that they "undertook with new confidence to repair the damages in the great capitals and the bases that supported the columns" after having joining "the stories of the old and new building." Abbot Suger, "Libellus alter de consecratione ecclasiae sancti Dionysii," *Abbot Suger on the Abbey Church of St.-Denis and Its Art Treasures*, trans. Erwin Panofsky (Princeton, N.J.: Princeton University Press, 1979), 95.

countless other medieval chroniclers.[11] Most material descriptions in Suger's text refer to gold, stained glass, and precious gems as well as the need to cover up the bare, unconcealed stones of the church. The stones themselves are not subject to these descriptions of opulence or beauty, but rather remarked upon because of their strength and sanctity.[12]

The abbot implies a certain sacredness and vitality in the stones of the church, referring to them as relics: "Deliberating under God's inspiration, we choose…to respect the very stones, sacred as they are, as though they were relics…"[13] Suger writes that the columns of his chevet are meant to represent the Twelve Apostles and, in the side aisles, the Minor Prophets, drawing another parallel between the stones of the church and their spiritual—and human—dimension.[14] In an intriguing reversal of the metaphor, the abbot compares visitors crowded into the interior of the church to "stone statues" because they were pressed up against one another, an image that evokes twelfth-century column figures whose limbs rarely escape the boundaries of a portal

---

[11] In Book II of the *Metamorphoses*, "materiam superabat opus" is used to describe the Palace of the Sun built from a multitude of precious materials: gold, bronze, ivory, and silver, as well as doors with engraved images of with cosmic scenes and adorned with the signs of the Zodiac (which also appear in the portal sculpture at Saint-Denis, see chapter 3, "Strange Weather"). The phrase was employed in many medieval texts, including Gervase of Canterbury, the Metrical Life of Saint Hugh, and Abbot Suger. See Suger, "Liber de rebus in administrione sua gestis," trans. Panofsky, 63.

[12] Suger emphasizes the need to cover unadorned stones with rich materials: "…we endeavored, God helping, to build [a tomb] very illustrious both by the exquisite industry of the goldsmiths' art and by a wealth of gold and precious stones. We made preparations to fortify it all round, outwardly noble for ornament by virtue of these and similar [precious materials], yet inwardly not ignoble for safety by virtue of a masonry of very strong stones; and on the exterior—lest the place be disfigured by the substance of unconcealed stones—to adorn it (yet not [so handsomely] as would be proper) with gilded panels of cast copper." Suger, "Libellus alter de consecratione," trans. Panofsky, 104-107.

[13] "Deliberando elegimus, ut propter eam quam divina operatio, sicut veneranda scripta testantur, propria et manuali extensione ecclesiae consecratione antiquae imposuit benedictem, ipsis sacratis lapidinus tanquam reliquiis deferremus…" Suger, "Libellus alter de consecratione," trans. Panofsky, 100-101.

[14] "The midst of the edifice, however, was suddenly raised aloft by columns representing the number of the Twelve Apostles and, secondarily, by as many columns in the side-aisles signifying the number of the [minor] Prophets, according to the Apostle who buildeth spiritually. *Now therefore ye are no more strangers and foreigners, says he, but fellow citizens with the saints and of the household of God; and are built upon the foundation of the apostles and prophets, Jesus Christ Himself being the chief cornerstone* which joins one wall to the other; *in Whom all the building*—whether spiritual or material—*groweth unto one holy temple in the Lord. In Whom we, too,* are taught *to be builded together for an habitation of God through the* Holy *Spirit* by ourselves in a spiritual way, the more loftily and fitly we strive to build in a material way." Suger, "Libellus alter de consecratione," trans. Panofsky, 104-105.

jamb, as well as the apostle-like columns supporting the Gothic chevet. While Suger has much to say about church decoration and the sanctity of the stones in the church, sculpted decoration is not part of his discourse.

The thirteenth-century liturgist William Durandus does not mention sculpted decoration specifically but also maintains a belief in the vitality and sanctity of the stones. He wrote that the church was constructed out "living stones" (*vivis ex lapidibus construitur*).[15] The stones, according to Durandus, represented the community of faithful individuals in the church, forever stacked on top of one another reaching up towards the heavens. Not all stones are equal in Durandus' eyes, however; the "bigger stones, and the polished or square ones that are placed on the outside wall—in the middle of which lie the smaller ones—are the more perfected men whose merits and prayers sustain the weaker men in the holy Church."[16] This metaphor has biblical precedents; in Isaiah 28:16, the Messiah is referred to as "a corner stone," and the Latin translation of St. Peter's letters expand on the metaphor to include the followers of Christ; Abbot Suger also describes episodes of holy men laying the cornerstones of Saint-Denis.[17] The importance of stone as a sanctified building material in churches is notable. It is worth considering how theological ideas about church decoration might have impacted how medieval viewers considered *sculpted* stone, as the vital qualities and liveliness of the material makes the incorporation of sculpted plant life an interesting confluence of material and subject matter.

---

[15] "Siquidem ecclesia materialis in qua populus ad laudandum Deum conuenit sanctam significat Ecclesiam que in celis vivis ex lapidibus construitur." William Durandus, *Rationale divinorum officiorum*, Book 1, 115-117. *Corpus Christianorum* vol. 140, p. 15.

[16] Durandus, trans. Thibodeau, p. 14.

[17] 1 Peter 2: 4-5: "Ad quem accedentes lapidem vivum, ab hominibus quidem reprobatum, A Deo autem electum et honorificatum: et ipsi tamquam lapides vivi superaedificamini, domus spiritualis sacerdotium sanctum offerre spiritales hostias acceptabiles Deo per Iesum Christum."

Durandus' exegesis on the meaning of the physical church and its furnishings draws on a long tradition of Latin literature. The concept of "living stone" has its roots in this tradition as well, likely lifted from the metaphor of *lapidis vivis* found in contemporary and ancient Latin texts. Descriptions of "living stones" appear in Virgil and Ovid, who used the term *vivum saxum* to refer to "living rock" rooted to or growing from the earth *in situ*.[18] The metaphor was employed by many Christian writers, including St. Augustine, the Venerable Bede, and medieval Christian poets. Durandus' particular phrasing makes use of the passive *construitur*, which also appears in the popular hymn *Urbs beata Ierusalem*, sung on the feast of the dedication:

> Blessed city of Jerusalem, called a vision of peace,
> Built in heaven out of living stone [Quae construitur in caelis *vivis ex lapidibus*],
> And crowned by angels like a bride for her spouse.[19]

The idea of living stone and the heavenly Jerusalem are linked in 1 Peter: 2: 1-8, in which living stone is likened to the body of the church and the congregation: "Unto whom coming, as to a living stone, rejected indeed by men but chosen and made honorable by God: / Be you also as living stones built up, a spiritual house, a holy priesthood, to offer up spiritual sacrifices, acceptable to God by Jesus Christ."[20] The attention to stones, the church, and the heavenly Jerusalem can also be found in Revelation 21:19-20 ("And the foundations of the wall of the city were adorned with all manner of precious stones…"), though the focus is on the preciousness of

---

[18] J. C. Plumpe, "Vivum Saxum, Vivi Lapides: The Concept of 'Living Stone' in Classical and Christian Antiquity," *Traditio* 1 (1943): 1-14.

[19] "Urbs beata Ierusalem, dicta pacis visio, / Quae construitur in caelis vivis ex lapidibus / Et angelis coronata ut sponsata comite." Qtd. in Plumpe, 12 (my translation). This hymn was revised by Pope Urban VIII in 1632 to include *viventibus saxis*. Plumpe posits this change was to better suit the meter.

[20] "ad quem accedentes lapidem vivum ab hominibus quidem reprobatum a Deo autem electum honorificatum / et ipsi tamquam lapides vivi superaedificamini domus spiritalis sacerdotium sanctum offerre spiritales hostias acceptabiles Deo per Iesum Christum" 1 Peter 2: 1-8, Latin Vulgate.

and variety of stones adorning the walls, rather than their vivacious qualities—a distinction also seen in the writings of Abbot Suger.[21]

The heavenly Jersusalem is built from living stone, but the idea of "living stone" is not only linked to the idea of organic plant life poetically. Stone, according to many medieval writers attentive to Ovid and biblical texts, took root in the earth and grew organically, just like any living plant, shrub, or tree.[22] To sculpt plant life in stone was not necessarily antithetical; these materials were not so unlike one another in their ability to generate from the earth. And yet, sculpted foliage could be seen to occupy a liminal space in regard to its materiality: this sculpture is neither dead nor living; it denies its identity as stone while often (though not always) falling short of identification as an individual plant.

In select Gothic buildings, sculpted foliage seems to petrify an interest in change and transformation into the very fabric of the church, as the sculpted plants installed into the sturdy architectural frame of the building guaranteed that the architecture itself would remain suspended in a permanent state of growth and regeneration. The sanctity of the building was made even more miraculous by the continuous presence of these stones, which appeared to grow out of the most architecturally significant parts of the building: flowers bloom from the boss stone of a

---

[21] "fundamenta muri civitatis omni lapide pretioso ornata fundamentum primum iaspis secundus sapphyrus tertius carcedonius quartus zmaragdus / quintus sardonix sextus sardinus septimus chrysolitus octavus berillus nonus topazius decimus chrysoprassus undecimus hyacinthus duodecimus amethistus" Revelation 21: 19-20, Latin Vulgate.

[22] Fabio Barry, "Painting in Stone: The Symbolism of Colored Marbles in the Visual Arts and Literature from Antiquity until the Enlightenment. Doctoral Dissertation, Columbia University, 2011. See especially 2-4; "'Living Rock' and Veining," 54-51. Barry explored the concept of "living rock" in his study of the symbolism of marble from antiquity to the Enlightenment, engaging deeply with written sources that describe rock as a living material that "grew and flowered." Barry highlights certain similarities in the way the material stone was likened to organically growing plants and animals with blood coursing through their veins. See also Fabio Barry, *Painting in Stone: Architecture and the Poetics of Marble from Antiquity to the Enlightenment* (New Haven ; London: Yale University Press, 2020), which I was unable to consult due to library restrictions during the Covid-19 pandemic.

quadripartite vault, or frame portals that demarcated the entrance into sacred space or run along vault ribs. Sculpted foliage could also implicitly serve to reinforce the power of the miraculous with the church, rather than out in the fields or in its surroundings. In this way, it could even be advantageous to carve idealized, non-realistic foliage, a point that will be explored in chapter four. The vegetal forms in the garland frieze in the interior of Notre-Dame of Amiens, or the exuberant capitals of Wells cannot be found in the surrounding countryside or cultivated fields; rather, these idealized forms belong only to the sacred space and religious authority of the cathedral. By staking a claim to miraculous change, cathedrals and churches that flowered forever would never face the corruption of death and decay. In this way, sculpting plant life into stone is of great importance: the permanence of the material, combined with the dynamism of generation and change, is a powerful, even poetic, combination. Like saints' bodies that were found intact and smelling sweetly, eternally flowering buildings exhibit, as Stephen Murray has described, an "explosive process of becoming,"[23] remaining in a constant state of grace, no matter the meteorological conditions of their surroundings.

In the following pages, I will examine key examples of foliate carving from the twelfth and thirteenth centuries, re-evaluating the historiographic context that has been used to understand foliate motifs and challenging ideas of architectural decoration, homogeneity, serialized production, and meaning in Gothic architecture during this period. The first chapter summarizes a series of historiographic issues concerning the production of sculpted foliage to frame the study within architectural history, considering the reception of Vitruvius, the role of Viollet-le-Duc in the restoration of foliate elements, as well as the status of iconography and ornament in medieval architecture. Looking to key medieval texts, the second chapter considers

---

[23] Stephen Murray, *Notre-Dame, Cathedral of Amiens: The Power of Change in Gothic.* Cambridge, New York: Cambridge University Press, 1996: 59.

natural philosophy in the twelfth and thirteenth centuries, examining how these texts intersect with interpretations of sculpted foliage and contemporary images of nature, positing that ecological change, rather than the influx of ideas to medieval universities, was a key factor in changing attitudes towards nature. The third chapter highlights the foliate program on the west façade of Notre-Dame of Amiens and its connections to local liturgical practice, authenticity of local relics, and subversion of natural phenomena in the cult of Saint Firmin, examining "environmental miracles" and depictions of weather. The fourth chapter takes one of the most significant examples of thirteenth-century foliate sculpture, the foliate frieze that runs the length of the interior of Amiens, applying a new analysis to argue that sculptors knowingly made this monumental frieze look as if it were organic and alive, rather than reproducing each segment serially. This chapter also includes the first full reproduction of the frieze to appear in print. Finally, the fifth chapter takes on the subject of crockets, examining the fluidity with which these modular architectural elements were applied and looking to a particular manifestation in the quire and on the west front of Wells Cathedral. This reimagining of foliate forms in Gothic architecture provides new avenues for studying and understanding this type of decoration, highlighting the utility of studying architectural decoration, especially in cases where few documentary sources exist.

1

Studying Gothic Foliage

Foliate sculpture has been a subject of inquiry for many historians of medieval art, cross-examined with a myriad of art historical methods and approaches. The existing literature, fortunately, is rich with insights; the ways in which Gothic foliage has been studied reflects the ever-evolving ways in which we study monuments of the past. Just as each generation "builds its own Gothic cathedral," so have analyses of foliate sculpture responded to the demands of each generation's inquiries into medieval sacred space, artistic production, and expression of architectural form.[24] What makes the historiography of foliate sculpture especially dynamic is the ability of this particular form to elude the somewhat limited and arbitrary categorization systems of the field: sculpture and architecture, structure and ornament, naturalism and abstraction, as well as standard iconographic themes common in the interpretation of Western medieval art.

The relationship between nature and Gothic architecture is a complex issue and has been recounted in detail by other art historians, some of whom have touched on issues relating to sculpted foliage. Much to the benefit of this study, Jean Givens and Paul Crossley have summarized trends in the historiography of sculpted foliage in Gothic architecture, highlighting

---

[24] Paul Crossley "Introduction," in Paul Frankl, *Gothic Architecture*, Rev. Ed. (London: Yale University Press, 2000), 30; Willibald Sauerländer, "Gothic: The Dream of an Un-Classical Style," in *Gothic Art and Thought in the Later Medieval Period: Essays in Honor of Willibald Sauerländer* (University Park, PA: Pennsylvania State University Press, 2011). In contextualizing the contributions of Paul Frankl, Paul Crossley identifies the interpretive challenges of studying Gothic architecture in a postmodernist climate in his historiographic preface, published in 2000. A decade later, Willibald Sauerländer published his historiographic essay suggesting that there were few paths that the conscionable art historian of Gothic architecture, lest they become nostalgic for romantic dreams of the un-Classical cathedral, could follow. Interestingly, Sauerländer concludes that art historians should pay more attention to liturgical texts to "transform the dream of the cathedral."

13

connections with romantic thought and trends in the practice of art history in Germany.[25]

Stephen Murray has explained the importance of natural origins in the narratives surrounding

Gothic architecture and the language that has defined this style of architecture, looking to

Vitruvius, Renaissance critics, James Hall, and others.[26] Mailan Doquang identified interpretive

themes in German and French scholarship, tracing the study of foliate sculpture in concert with

the dynamics of the evolving relationship between structure and ornament in architectural

history, with debts to the field of Islamic architecture and, in particular, the work of Barry

Flood.[27] Building on the rich historiographic studies of these scholars as well as those who have

contributed insights into the historiography of Gothic architecture more broadly, my analysis

explores how architectural foliate sculpture has been rendered into written descriptions in

architectural treatises and critical texts.

This chapter begins with the observations of Vitruvius and how his theories informed

medieval thinkers' ideas about architectural space, sculpted ornament, and the relationship

between architecture and nature. In the Italian Renaissance, however, this same text was used to

question and even malign the propriety of Gothic structure and its sculpted, foliate ornament, a

rhetorical pattern that was repeated by architectural critics in eighteenth-century England.

Viollet-le-Duc's monumental project, the *Dictionnaire raisonné de l'architecture française*, not

only represents a turn in the study of medieval architecture, but an attempt to present an updated

form of the architectural treatise—one that happens to include extensive illustrations of and

---

[25] Jean Givens, "Gothic Naturalism" in *Observation and Image-Making in Gothic Art* (New York: Cambridge University Press, 2005), 5-36; Paul Crossley, "The Return to the Forest: Natural Architecture, the German Past in the Age of Dürer," in *Künstlerischer Austausch: Akten des XXVIII Internationalen Kongresses für Kunstgeschichte, Berlin, 15-20 juli, 1992*, ed. Thomas W. Gaehtgens, vol. 2 (Berlin: Akademie Verlag, 1992), 71–80.
[26] Stephen Murray, *Plotting Gothic* (Chicago: University of Chicago Press, 2014), 138-148.
[27] Mailan S. Doquang, *The Lithic Garden: Nature and the Transformation of the Medieval Church* (New York, NY: Oxford University Press, 2018), 9-19.

reflections on foliate sculpture, playing a large role in shaping the discourse around sculpted

foliage up to the middle decades of the twentieth century. While this project differed widely from

the architectural criticism of John Ruskin, he and Viollet-le-Duc agree on some aspects of foliate

decoration—perhaps both writers, when theorizing about the relationship between art and nature,

were informed by similar strains of romanticism in England and France. Finally, I will explore

art historians' attempts to find meaning in foliate sculpture that extend beyond the rhapsodies of

romanticism, using the tools of iconography and visual analysis were employed in the Émile

Mâle's influential volumes on medieval art. In reviewing how these tools fell short in previous

generations of scholarship, it becomes clear that there are many opportunities for new

frameworks in the study of sculpted decoration—and architectural decoration more broadly.

1.1 Vitruvius, Gender, and the Reception of Gothic

First, it is useful to review how Vitruvius included discussions of vegetal ornament and

architectural decoration within the context of his treatise. Vitruvius comments on sculpted

foliage within the broader context of architectural decoration as well as in his explanation of the

varying quality and geographies of stone quarries. Nature, writ large, was central to the

understanding of architecture in his treatise; architects, he argued, must have a deep

understanding of the machinations of nature in order to plan and build and plan properly,

including having knowledge of meteorology, geometry, astronomy, the particulars of diverse

climates as well as the materiality of diverse building materials. According to Vitruvius, the

quality of stone had a direct effect on the quality of the carved ornament and its ability to

weather the elements: he writes that the close-textured stone from the area around Ferento and

the Lago da Bolsena "cannot be injured by the weather or force of fire," which was a reason why the old monuments in this area, among them "large statues exceedingly well made, images of smaller size, and flowers and acanthus leaves gracefully carved…look as fresh as if they were only just finished."[28] If Rome had access to quarries like these, he writes, the buildings would be in far better condition. High-quality stone of a certain density was ideal for sculpting delicate floral decoration and keeping buildings and statuary in good condition, more suitable to weather the ravages of time or adverse conditions.

Several passages concerning the proper use of sculpted foliage can be found in Book 4 of the treatise, which explains and defines the Classical orders. It bears mentioning that, according to Vitruvius, the Doric order took its proportions from the measurements and *venustas* of men, while the Ionic order took its slender proportions and decorative details from women, *mulieres*—the Corinthian order, however, was associated with *virgines:* young, unmarried maidens. The Corinthian order was a melding of the other two but also distinguished by the foliate carving on its capitals; the delicacy of this carving was suitable for its slender, maiden-like proportions. The evocative origin story of the Corinthian order, as recounted by Vitruvius, further underscores the relationship between virginal maidens, foliage, and delicacy in architectural design. The artisan Callimachus, having discovered acanthus leaves growing over an offering placed upon the tomb of a young maiden in Corinth, was inspired to combine foliate carving with slender columns, thus creating a new architectural order. The most decorative and overtly vegetal of the orders defined in this treatise is associated not only with youthful proportions, but with death, offering, and remembrance.

---

[28] All English translations of Vitruvius are from Morris Hicky Morgan, whose translation was based on Valentin Rose's Latin edition of 1867. Latin terms used in this text are from Rose's volume. Vitruvius, *The Ten Books on Architecture [De architectura]*, Book 1, trans. Morris Hicky Morgan (New York: Dover Publications, 1960), 50; Vitruvius, Hermann Müller-Strübing, and Valentin Rose, *De architectura: libri decem* (Lipsia: Teubner, 1867).

In applying sculpted foliage to architectural design, Vitruvius emphasized the importance

of decorum, or propriety, which he defines as the "perfection of style which comes when a work

is authoritatively constructed on approved principles" which "arises from prescription, from

usage, or from nature."[29] The Doric order was suitable for temples dedicated to Minerva, Mars,

and Hercules—any warrior deity, regardless of gender, "since the virile strength of these gods

makes daintiness entirely inappropriate to their houses."[30] Temples to Venus, Flora, Nymphs and

other "delicate divinities" should employ the Corinthian order, so that "its rather slender outlines,

its flowers, leaves, and ornamental volutes will lend propriety where it is due."[31] Vitruvius does

not malign floral or vegetal architectural decoration, nor does he apply an aesthetic judgment to

its usage; rather, he includes guidelines for how he believes this type of architectural decoration

should be used, choosing to associate it with the delicacy of young maidens, nymphs, and certain

goddesses.

Vitruvius' explanations of the role of gender in architectural form is not limited to

masculine and feminine; he also includes age and class.[32] The distinction between *mulieres* and

virgines illustrates two distinct shades of Vitruvian feminine propriety in architectural

expression. This point is also evident in his discussion of Caryatid figures, in which Vitruvius

recalls the sacking of Caryae, the Peloponnesian city that sided with the Persians against the

Greeks. The Greek soldiers, he wrote,

> took the town, killed the men, abandoned the State to desolation, and carried off their
> wives into slavery, without permitting them, however, to lay aside the long robes and
> other marks of their rank as married women, so that they might be obliged not only to

---

[29] Vitruvius, trans. Morgan, Book 1, Ch. 2, 114-15

[30] Vitruvius, trans. Morgan, Book 1, Ch. 2, 115

[31] Vitruvius, trans. Morgan, Book 1, Ch. 2, 115.

[32] Matthew M. Reeve, *Gothic Architecture and Sexuality in the Circle of Horace Walpole* (Penn State
University Press, 2020), esp. 1–18. Reeve's analysis of the reception of Vitruvius and Gothic architecture in
the circle of Horace Walpole in the context of eighteenth-century England is especially illuminating on this
point.

march in the triumph but to appear forever after as a type of slavery, burdened with the weight of their shame and so making atonement for their State.[33]

These captured women, commemorated in the architectural form as load-bearing column figures, blended their status as married and enslaved, keeping their garments and insignia that designated them as *mulieres*. This is all to say that the status of a female person in Vitruvius' text, real or imagined, is malleable and highly contextual, based on marital status, age, and even place of origin. Pregnancy also alters a woman's status and value: in a passage about felling timber in the autumn, Vitruvius compares spring trees to pregnant women, whose bodies "are not considered perfectly healthy until the child is born," which is why

> pregnant slaves, when offered for sale, are not warranted sound, because the fetus as it grows within the body takes to itself as nourishment all the best quality of the mother's food, and so the stronger it becomes as the full time for birth approaches, the less compact it allows that body to be from which it is produced.[34]

The complex aspects of gender in Vitruvius' treatise were, of course, filtered through his own sociocultural context; nonetheless, the alignment of sculpted foliage with the feminine is striking, especially in light of how Renaissance theorists who had studied *De architectura* characterized the sculpted foliage of Gothic buildings, a point to which we will return.

1.1.1 Textual Transmission of Vitruvius in the Middle Ages and Renaissance

While the "discovery" of Vitruvius was lauded by a particular circle of Italian Renaissance humanists—and this myth persevered in certain threads of art historical scholarship—the treatise had been circulated and read for centuries prior to the first printed edition published in 1485. *De architectura* was conserved in a number of medieval scriptoria in France, England, Germany and the Low Countries in the Middle Ages, with the earliest existent

---

[33] Vitruvius, trans. Morgan, Book 1, Ch. 1, p. 6.
[34] Vitruvius, trans. Morgan, Book 2, Ch. 9. p. 58

manuscripts dating to the Carolingian period.[35] Of the 78 known Vitruvian manuscripts dating

from the eighth to the fifteenth centuries, 48 are dated before 1400. The earlier manuscripts often

contain the complete text combined with other practical treatises, such as Palladius' *De*

argricultura and Vegetius' *De re militari*—though other versions are conserved in fragments.

Vitruvius' text, it appears, was one of many practical treatises on worldly concerns available to

medieval scholars—and it is clear from medieval literature that many scholars knew Vitruvius'

work. Thierry de Saint Trond names Vitruvius in two poems and *De architectura* is mentioned in

the work of Hugh of St. Victor, while Vincent of Beauvais included Vitruvius' passage on the

proportions of the human body into his encyclopedic work *Speculum maius* and Albertus

Magnus seems to have also known the architecture treatise well. Carole Krinsky provided

evidence that *De architectura* was conserved in the libraries of Rouen and Cluny in the twelfth

century; the medieval manuscripts originated from production centers in Germany, northern

France, and England beginning as early as the ninth century, with several surviving copies

created in the twelfth and thirteenth centuries.[36]

---

[35] Edgar de Bruyne traced the impact of Vitruvius in medieval Europe, while catalogues of existent
manuscripts have been documented by Carol Herselle Krinsky and Stefan Schuler; L.D. Reynolds and S.F.
Weiskittel defined distinct groups within existent manuscripts and dissemination within manuscript production
centers and collections. Wim Verbaal placed these studies within historiographic context. Architectural
historians have looked for evidence of Vitruvian ideas in medieval monuments, such as Kenneth Conant's
study of Cluny III, though it can be difficult to prove that building measurements and proportions share a
direct connection with *De architectura*. Peter Kidson also summarized the transmission of the text in the
Middle Ages and its possible influence on medieval architectural production; see Kidson, "Vitruvius," in *The
Dictionary of Art*, ed. Jane Turner, vol. 32, 34 vols. (New York: Grove, 1996), 632–42. Edgar de Bruyne,
Études d'esthétique médiévale, vol. 1 (of 3), Rijksuniversiteit te Gent. Werken uitg. door de Faculteit van de
Wijsbegeerte en Letteren,(Brugge: De Tempel, 1946), 234-255; Carol Herselle Krinsky, "Seventy-Eight
Vitruvius Manuscripts," *Journal of the Warburg and Courtauld Institutes* 30 (1967): 36–70; Stefan Schuler,
Vitruv im Mittelalter : die Rezeption von "De architectura" von der Antike bis in die frühe Neuzeit (Köln:
Böhlau, 1999); L. D. Reynolds and S. F. Weiskittel, "Vitruvius," in *Texts and Transmission: A Survey of the
Latin Classics*, ed. L. D. Reynolds and R. A. B. Mynors (Oxford, New York: Clarendon Press ; Oxford
University Press, 1983), 440–45; Wim Verbaal, "The Vitruvian Middle Ages and Beyond," *Arethusa* 49, no. 2
(Spring 2016): 215–25; Kenneth J. Conant, "The After-Life of Vitruvius in the Middle Ages," *Journal of the
Society of Architectural Historians* 27, no. 1 (1968): 33–38.
[36] Reynolds and Weiskittel, "Vitruvius," 441-443. Many of the manuscripts are similar to BL Harley 2767, a
ninth-century manuscript that might have originated from the scriptorium of Charlemagne and copied at

In Italy, however, it seems that the text was not widely circulated before the first printed edition. Though surviving manuscripts suggest that *De architectura* was likely conserved in the library of Montecassino, the reception of the text in Italian literary circles, at least in part, can likely be traced to Petrarch's copy of the text, which he acquired in France in the mid-fourteenth century.[37] The transmission of *De architectura* expanded widely after the text was printed, allowing a variety of patrons across Europe to add Vitruvius to their growing book collections, especially in private, non-monastic settings. Despite the text's outsized presence after the first printed edition of 1485, it is important to acknowledge that separate strands of readership and transmission occurred quite outside the context of the Italian Renaissance.

Separating these instances of textual transmission is particularly useful in considering historiographical questions for medieval art history; many generations of scholarship were predicated on certain assumptions built into the foundational texts of the discipline and their interpretation on the part of Italian Renaissance theorists. That said, these texts have also shaped the discourse of architectural theory, architectural decoration, and the idea of decorum or propriety in architectural expression more broadly, especially in context where Classical or neoclassical modes of building are prioritized, as in the Italian Renaissance, vogue for Palladian

---

Corbie. It is thought that the text traveled over the Channel in the tenth century. The surviving texts populated scriptoria and libraries in France, England, the Low Countries, and England; while medieval copies of Vitruvius were kept at Montecassino.

[37] Reynolds and Weiskittel, 443; Peter Fane-Saunders, "The Manuscript Hunter and the Librarian: Poggio Bracciolini and Giovanni Tortelli," in *Pliny the Elder and the Emergence of Renaissance Architecture* (New York, NY: Cambridge University Press, 2016), note 4; Poggio Bracciolini and Niccolò Niccoli, *Two Renaissance Book Hunters: The Letters of Poggius Bracciolini to Nicolaus de Niccolis ; Translated from the Latin and Annotated*, electronic resource, trans. Phyllis Walter Goodhart Gordan, Records of Western Civilization (New York: Columbia University Press, 1991), 191. Petrarch was the teacher of Franciscus de Fiana, whose pupil Cincius Romanus penned the letter describing the "discovery" of Vitruvius' *De architectura* in the library of St. Gall. It is possible that this pedagogical lineage included the instruction of the text, priming Romanus' eyes to "well with tears" when he came across the dusty medieval manuscript in Germany.

villas in England and abroad, as well as the architectural education at the École des Beaux-Arts and academies like it.

1.1.2 Leaves "beyond all natural reason"

When Poggio Bracciolini found a manuscript containing Vitruvius' *De architectura* in the library of St. Gall in the mid 1410s, his companion Cincius Romanus documented their discovery in a letter to his teacher Franciscus de Fiana, a former pupil of Petrarch, along with his thoughts on the condition of the monastic library and its conservators.

> In Germany there are many monasteries with libraries full of Latin books. This aroused the hope in me that some of the works of Cicero, Varro, Livy, and other great men of learning, which seem to have completely vanished, might come to light, if a careful search were instituted. A few days ago, Poggius and Bartholomeus Montepolitanus and I, attracted by the fame of the library, went by agreement to the town of St. Gall. ... But when we carefully inspected the nearby tower of the church of St. Gall in which countless books were kept like captives and the library neglected and infested with dust, worms, soot, and all things associated with the destruction of books, we all burst into tears, thinking that this was the way in which the Latin language had lost its greatest glory and distinction. Truly, if this library could speak for itself, it would cry loudly: "You men who love the Latin tongue, let me from this prison in whose gloom even the bright light of books within cannot be seen." There were in that monastery an abbot and monks totally devoid of any knowledge of literature. What barbarous hostility to the Latin tongue! What damned dregs of humanity![38]

Romanus goes on to lament the loss of the libraries of Rome, express his sense of injustice at the destruction of texts, and condemn what he saw as the German monks' disrespect towards the collections of Latin texts in their aging scriptoria. Indeed, as he says in his letter, he and his colleagues shared an outburst of emotion in the "captive" library of St. Gall. Their mission to "liberate" these texts from their dark and dusty confines speaks volumes of how Italian

---

[38] Cincius Romanus, "Cincius Romanus to his most Learned Teacher Franciscus de Fiana," in *Two Renaissance Book Hunters: The Letters of Poggius Bracciolini to Nicolaus de Niccolis; Translated from the Latin and Annotated*, trans. Phyllis Walter Goodhart Gordan, Records of Western Civilization (New York: Columbia University Press, 1991), 187-191.

humanists saw themselves in relationship to the foreign guardians of these manuscript collections, obscuring the pre-Renaissance transmission of the texts.

In the writing of Vasari and Raphael, famously among the first generation of writers to figure "Gothic" architecture as stylistically divergent from the buildings of their own sociocultural context, "German" buildings are described not unlike the state of the dusty library of St. Gall: irrational, unlearned, even barbaric. Furthermore, foliate decoration was seen as a sign of structural and even aesthetic weakness, a symptom of the architects' lack of learning and application of propriety. Raphael's letter to Pope Leo X has been quoted often by scholars of Gothic architecture, but it is useful to return to this passage to consider how foliate decoration plays an important role in his critique.[39]

> ...the Germans (whose style still endures in many places) for ornament often placed just some little hunched, ill-formed figure, as a bracket to support a beam, or strange animals, and figures, and coarse foliage, beyond all natural reason. However their architecture had this origin, that it was born from trees not yet cut down, which, branches bent and tied together, make up their pointed arch; and although this origin may not be entirely to be scorned, still, it is weak; for shelters composed of beams fastened together, and posts used as columns, with ridges and coverings, as Vitruvius writes of the origin of Doric building, would support much more than the pointed arches, which have two centers.[40]

Notably, Raphael describes the German/Gothic architectural decoration as beyond all natural reason (*fuori d'ogni ragione naturale)*. It is the "ill-formed" and "coarse" decorative forms are the subject of the author's ire, rather than the buildings' structural systems or the allusion to the natural origins of architecture, which he relates to Vitruvian theories and may not be entirely "scorned." While the structural system has its flaws, it is the architectural decoration that is beyond Raphael's idea of decorum.

---

[39] Doquang, *The Lithic Garden,* 13; Murray, *Plotting Gothic,* 139; Ingrid Rowland, "Raphael, Angelo Colocci and the Genesis of the Architectural Orders." *The Culture of the High Renaissance,* 228-30; Paul Frankl, "The Attitude of the High Renaissance" in *The Gothic,* 261-284, esp. 271-278.

[40] Raphael, "Letter to Pope Leo X," in Carlo Pedretti, *A Chronology of Leonardo Da Vinci's Architectural Studies after 1500,* Travaux d'humanisme et Renaissance 54 (Genève: E. Droz, 1962), 165.

Vasari's comments on Gothic architecture appear in the third chapter of

Dell'Architettura, to which a discussion of the German architecture was appended to a

discussion of the five orders (*De'cinque ordini d'archittetura, rustica, dorico, ionico, corinto,*

composto, e del lavoro tedesco).[41] Vasari foregrounds his comments on Gothic architecture with

his discussion of Classical norms. The oft-cited passage on the *lavori tedeschi* appears at the end

of the chapter as a contrasting addendum and a warning: not only was Gothic architecture

structurally weak, he argued, it was taking over Europe like an invasive species, not unlike the

twisting vines and leaves spreading over its surfaces, threatening collapse.[42] Vasari writes:

> There is another type of works called *tedeschi*, which in their ornaments and proportions
> are very different from the antique and the modern. Today they are not used by the most
> gifted architects, who instead flee from them as monstrous and barbarous and forsaken of
> all that comprises order—which should rather be called confusion and disorder—for they
> have made in their buildings, which are so numerous that they have polluted the world,
> portals adorned with columns that are thin and twisted like vines [*colonne sottili ed*
> *attorte a uso di vite*] and do not have the strength to bear a load, no matter how light. And
> so, for all the surfaces and their other ornaments, they made a curse of tabernacles [*una*
> *maledizione di tabernacolini*], one on top of the other, with so many pyramids and points
> and leaves [*piramidi e punte e foglie*], that it seemed impossible that it might stand...[43]

---

[41] Vasari, Giorgio. *Le vite de' piu eccellenti pittori, scultori, e architetti*. Edited by Gaetano Milanesi, Carlo
Milanesi, Carlo Pini, and Vincenzo Fortunato Marchese. Vol. 1 (of 13). Raccolta Artistica. Firenze: Felice Le
Monnier, 1846, 114-122.

[42] Murray, *Plotting Gothic*, 140; Anne-Marie Sankovitch, "The Myth of the 'Myth of the Medieval': Gothic
Architecture in Vasari's Rinascita and Panofsky's Renaissance," *Res: Anthropology and Aesthetics* 40
(September 2001): 34. See also Paul Frankl, *The Gothic*, 290-91; Erwin Panofsky, *Meaning in the Visual Arts*,
188-190. Murray and Sankovitch highlighted the threatening, geographical "spread" of Gothic architecture and
its persistence. Sankovitch also examined this description in comparison to the classical orders. This passage
has been examined numerous times and read in a number of ways.

[43] "Ecci un'altra specie di lavori che si chiamano tedeschi, i quali sono di ornamenti e di proporzione molto
differenti dagli antichi e dai moderni. Né oggi s'usano per gli eccelenti, ma son fuggiti da loro come mostruosi
e barbari, mancando ogni lor cosa di ordine; che più tosto confusione o disordine si può chiamare, avendo fatto
nelle lor fabbriche, che son tante che hanno ammorbato il mondo, le porte ornate di colonne sottili ed attorte a
uso di vite, le quali non possono aver forza a reggere il peso, di che leggerezza si sia. E cosi, per tutte le facce
ed altri loro ornamenti, facevano una maledizione di tabernacolini l'un sopra l'altro, con tante piramidi e punte
e foglie, che, non ch'elle possano stare..." Vasari, Giorgio, Gaetano Milanesi, Carlo Milanesi, Carlo Pini,
Vincenzo Fortunato Marchese, and Lorenzo Ghiberti. *Le Vite de' Piu Eccellenti Pittori, Scultori, e Architetti*.
Vol. 1. Raccolta Artistica. Firenze: Felice Le Monnier, 1846, 122.

In addition to the striking turns of phrase in Vasari's text, it appears that his concept of German architecture is contemporary, not placed within a forgotten past like the neglected volumes in the St. Gall library, awaiting rescue. Rather, this style of architectural decoration—with its columns "twisted like vines" and leaves sprouting out of its surfaces—threatened, to Vasari, not only the structural integrity of buildings, but also a type of cultural collapse, an ideological sacking of Rome at the hands of the *Goti*.

But the tension between Italian and German architectural modes following the availability of the first printed edition of *De architectura* was, in fact, multi-directional. Connecting the availability of Vitruvius to the German humanist patron Wilhelm von Reichenau, Bishop of Eichstätt, the dedicatee and presumed patron of Matthäus Roriczer's treatise on the geometry of finials and pinnacles, Crossley convincingly pointed to the availability of printed editions of Vitruvius and Alberti as motivating forces in codifying, in written form, theories of Gothic architectural design.[44] Roriczer's treatise, the first of its kind to address Gothic architecture in this theoretical mode, was published in 1486, one year after the first printing of Vasari and Alberti's texts. Crossley argues that the treatise is an attempt to reconcile the natural origins and foliate design of German Gothic architecture with the popular consumption of humanist texts on architectural theory. Perhaps most importantly, the appearance of this treatise seemed to signal the emergence of a new audience for architectural theory: the educated patron. Roriczer's text takes on architectural elements that blend foliate forms and dynamic geometry, two forces that had been at play for centuries in Gothic design, as seen through the use of crockets, trefoils, quatrefoils, foliate friezes, and multi-petaled rose windows.[45]

---

[44] Crossley, "Return to the Forest," 74.

[45] An English translation of the treatises can be found in Matthäus Roriczer and Hanns Schmuttermayer, *Gothic Design Techniques: The Fifteenth-Century Design Booklets of Mathes Roriczer and Hanns Schmuttermayer*, trans. Lon Royce Shelby (Carbondale: Southern Illinois University Press, 1977); For a

When Vasari and Raphael wrote about German buildings, both writers seem to refer to a generalized, fictive ideal of Gothic architecture, an unknown group of buildings that likely included many examples that were approximately contemporary to the author's day. Neither names a specific building, nor a specific region. Gothic art and architecture of the late fourteenth and early fifteenth centuries, especially in Germany, displayed an extraordinary virtuosity in foliate carving in stone and in wood, a phenomenon that has been documented in a number of valuable and diverse art historical studies.[46] It is worth considering whether these were the buildings they held in their mind's eye as they wrote their diatribes, or whether they drew distinctions between Gothic churches of their contemporaries and those built centuries prior. It is clear from Vasari's anxiety about the spread of Gothic architecture, for example, that he was not placing *lavori tedeschi* in a distant, medieval past; this was the architectural production of his day, concurrent with the developments of Italian Renaissance architectural theory and practice.

Humanists' distrust of the Gothic structural system echoes Bracciolini's distrust of the monks in charge of the library at St. Gall, even though the geometrical proportion at the heart of Gothic design shared characteristics in common with Vitruvian ideals of symmetry, geometry, and perhaps even propriety. In later iterations of neoclassical architecture, experiments in Gothic design would not only be located squarely outside of Vitruvian and Palladian ideals of decorum

---

facsimile of the treatise, see Matthäus Roriczer, *Matthäus Roriczer: Baukunst Lehrbuch*, ed. Wolfgang Strohmayer (Hürtgenwald: G. Pressler, 2009).

[46] See, for example, Gregory C. Bryda, "The Exuding Wood of the Cross at Isenheim,"; Hubertus Günther, "Das Astwerk und die Theorie der Renaissance von der Entstehung der Architektur," in *Théorie des arts et création artistique dans l'Europe du Nord du XVIe au début du XVIIIe siècle*, ed. Michèle-Caroline Heck, Collection UL3 Travaux et recherches (Colloque international, Villeneuve d'Ascq (Nord): Université Charles-de-Gaulle – Lille 3, 2001), 13-32; Étienne Hamon, "Le naturalisme dans l'architecture française autour de 1500," in *Le gothique de la Renaissance: actes des quatrième Rencontre d'architecture européenne, Paris, 12-16 juin 2007*, ed. Monique Chatenet et al., De architectura. Colloques 13 (Rencontres d'architecture européenne, Paris: Picard, 2011), 329–43; Ethan Matt Kavaler, "On Vegetal Imagery in Renaissance Gothic," in *Le gothique de la Renaissance*, ed. Monique Chatenet et al., 298-312 and "Natural Forms," in *Renaissance Gothic* (2012), esp. 201-220.

and propriety, but employed by a circle of patrons who found themselves quite outside the

traditional gender roles of eighteenth-century Britain. Matthew Reeve recently framed the

architecture of Strawberry Hill and satellite houses of Gothic design in England (Grove House,

Lee Priory, Donnington Grove, as well as a residence aptly named the Vyne) as "products of

homoerotic culture" resistant to dominant neoclassical trends, adding that "sexuality haunts the

reception history of the Gothic style like a ghost in its shell."[47] Gender, too, haunts the discourse

around architectural decoration and foliate ornament in particular; the transformations of gender,

propriety, and architectural design are fluid and dependent on a multi-faceted network of cultural

factors. While Enlightenment humanists would later interpret the classical orders in terms of a

male-female gender binary, Vitruvius' introduction of the myths of Callimachus as well as the

sack of Caryae illustrate that cultural conceptions of gender play a foundational role in

architectural criticism, particularly in Western narratives that assign propriety to certain modes

of architectural decoration.

1.1.3 Gender, Foliage, and Gothic Structure

When we look at Vitruvius in relation to medieval sculpted foliage, it is worthwhile to try

to disentangle Vitruvian ideas from their Renaissance and Enlightenment interpretations. Tracing

these developments complicates and enriches the understanding of humanist dismissals of Gothic

architecture; it also highlights how humanist critics, whether writing from the perspective of the

Renaissance or the Enlightenment, signposted Gothic vegetal ornament as a departure from how

they defined proper architectural expression. The critiques of Gothic architecture's weakness and

overwrought foliage draw a direct contrast to the values of strength, order, and propriety. It is

---

[47] Matthew M. Reeve, *Gothic Architecture and Sexuality in the Circle of Horace Walpole*, 2, 3.

worth considering whether the delicate, foliate, slender—and dare we say, feminine—aspects of Gothic design were occasionally invoked to cast this system of architectural expression, and those who employed it, as "unlearned," a coded word for "other." Furthermore, it is notable that a generation of scholars sought to rescue the rationalism and engineering ingenuity of the Gothic *structure*, often at the cost of its special decorative qualities. The monumental foliate band at Notre-Dame at Amiens, for example—for which no comparable example of medieval carving exists—was seen by some nineteenth-century critics as a rather embarrassing regression, interrupting the vertical, upward thrust of the structural piers and engaged shafts.[48] Building structural models and studying flying buttresses yields valuable information about these buildings—so, too, however, does the study of architectural decoration.

Vasari, Raphael, and even Poggio Bracciolini and his colleagues played large roles in shaping the discipline of art history, its organization, and how medieval art—especially Gothic architecture, which maintains its ties to barbaric roots through its very name—was placed into context of art historical narratives. There is still much to recover and attempt to understand in the architectural expression of the twelfth and thirteenth centuries, particularly in foliate carving: the flexibility of its decorative systems, its ability to heighten architectural form, the way it could suggest meaning, its activation through liturgical activities, and even its fabrication and design.

1.2 Viollet-le-Duc: Drawing and Replacing Foliate Sculpture

The writings and illustrations of Eugène-Emmanuel Viollet-le-Duc were instrumental in recovering not only Gothic architecture, but foliate sculpture, placing the foliate decoration of

---

[48] See footnote in Eugène-Emmanuel Viollet-le-Duc, "Bandeau" in *Dictionnaire raisonné de l'architecture Française du XIe au XVIe siècle*, vol 2 (Paris: Ve. A. Morel, 1874), 108-110.

Gothic architecture within a new theoretical framework. In his multivolume *Dictionnaire raisonné de l'architecture française du XIe au XVIe siècle,* written in several installments between 1857-1868, Viollet-le-Duc proposed a new format for engaging with architectural theory: his *Dictionnaire,* following conventions of encyclopedic reference books popularized by the publishing houses of nineteenth-century France, separated elements of architectural design and history into alphabetized entries, a departure from categorized hierarchies and exegeses on the classical orders. Viollet-le-Duc's principal subject of inquiry was French medieval architecture, a sustained focus that reflected his work as a restorer of medieval French monuments as well as his rejection of the École des Beaux-Arts. His *Dictionnaire,* notably, included thousands of engravings made from his own drawings—3367 illustrations in total.[49] Within Viollet-le-Duc's original surviving drawings, many of the original sketches for these illustrations are small in size and executed during the years of his restoration campaigns, suggesting that Viollet-le-Duc drew from observations on-site. Furthermore, many of the drawings address minute details of decorative carving in Gothic monuments, much of it foliate. Because Viollet-le-Duc and his teams were charged with replacing, re-fabricating, or—in some instances—redesigning these foliate elements altogether, he was uniquely knowledgeable about and deeply engaged with Gothic foliate carving.

As a restorer, Viollet-le-Duc documented foliate sculpture extensively in his working drawings. He drew studies of surviving passages of medieval sculpture, fabricated new patterns for replacement to be executed by his team of sculptors, and sometimes depicted passages of sculpture that were in poor condition at the time of restoration, showing erosion and damage

---

[49] Barry Bergdoll, "The Dictionnaire Raisonné: Viollet-Le-Duc's Encyclopedic Structure for Architecture," in *The Foundations of Architecture: Selections from the Dictionnaire Raisonné,* by Eugène-Emmanuel Viollet-le-Duc, trans. Kenneth D. Whitehead, First edition (New York: G. Braziller, 1990), 2.

brought about by the aging composition of the limestone and its delicate fabrication. These

drawings of damaged architectural sculpture were typically not chosen for publication in the

Dictionnaire, however, which depicts a wide range of architectural diagrams and samples of

sculpted keystones, capitals, and foliate bands. This approach would draw a contrast with the

later writings of John Ruskin, who customarily included illustrations of buildings' degradation

and included incisive critiques of restoration efforts. Nonetheless, many of Viollet-le-Duc's

published engravings document examples of foliate ornament, drawing attention to these minute

details by eliminating them from their architectural context while providing the viewer with

schematic information about size, moldings, or even how they interact with other architectural

elements of the building.

One reason for Viollet-le-Duc's focus on foliate details was that the delicacy of this type

of sculpture made it even more vulnerable to degradation over time and in need of restoration

efforts, a point that was not lost on the architect. Deeply undercut vines "freed" from the

constraints of solid masonry as well as crockets and finials that projected into space required

special care and frequent replacement. Such sculptural details naturally erode or break over

time—a limitation even commented upon by Vitruvius in his comments on the dense stone near

Ferento—furthermore, these vulnerable bits of sculpture could be easily damaged in significant

meteorological events. Even during the course of restoration, early spring storms were likely to

damage the freshly executed work on these vegetal forms. In a note of 1869, Viollet-le-Duc

recounted a storm that occurred on the 12th of February in Amiens, knocking down a stone

fleuron from the west façade of the cathedral, which landed on top of the roof of a nearby house

("*sans la défoncer,*" he notes, perhaps to soften the blow). This storm also damaged two

pyramidions in the King's Gallery, took down a *clocheton*, and threatened the mullions of the south transept windows such that the transept had to be closed to the public. The repairs were not difficult to complete but were a source of concern for the budget (*"ils ont une certaine gravité au point de vue de la dépense"*)[50]. In 1876, another spring storm damaged the cathedral roof as well as many pinnacles; one even broke and fell on the aisle roof.[51]

Vegetal sculpture was expensive to fabricate and replace. In the invoices from the Duthoit brothers, local sculptors hired to execute Viollet-le-Duc's vision for the restoration of Notre-Dame of Amiens, these foliate details are itemized: for the north tower, 120 crockets for the *grande pyramide* priced at 780 francs, 24 crockets for the *pyramidions,* 108 francs. Ten *fleurons* and ten foliate capitals.[52] For the portals, the Duthoit brothers itemized 68 crockets, the *bouquets* crowning the gables of the north and south portals, *rosaces* and *fleurons, feuillage,* and 22 meters of the *frise de roses aux arcades des grand porches,* the delicate rose frieze that had been heavily damaged and documented carefully in drawings by the architect. In 2021 US dollars, the cost of each individual crocket works out to be approximately $109.00, a hefty sum considering the amount of these units that cover so many surfaces of the building's exterior; the crockets on the *grand pyramid* alone cost the approximate 2021 equivalent of $13,045.00.[53]

---

[50] "Note de Viollet-le-Duc sur les dégât causés par la tempête du 12 février." 1869. Private archives of Genviève Viollet-le-Duc, now conserved in Paris at the Médiathèque de l'architecture et du patrimoine. Also transcribed in Michel Barjon, ed., "La cathédrale d'Amiens façade occidental: les restaurations des XIX et XX siècles" (Direction régionale des affaires culturelles de Picardie, Conservation régionale des monuments historiques, Groupe de Recherche Art Histoire Architecture et Littérature, Paris, 1993), Bibliothèque Louis Aragon, Amiens.

[51] "1876, 12 mars. Lettre du Préfet de la Somme au Ministre." Archives départementales de la Somme, V 431.108.

[52] "Travaux de sculpture exécutés par les Duthoit durant l'année 1861. Cathédrale d'Amiens. Travaux de sculpture exécutés par ordre et sur les dessins de M. Viollet-le-Duc, architecte, par Duthoit frères à Amiens." Archives privées Duthoit-Ansart, Extrait du volume "1860-1862. Salaires et devis" Also transcribed in Barjon, ed., 1993, Bibliothèque Louis Aragon, Amiens.

[53] Approximation based on the valuation of grams of gold in French francs in 1869, compared with valuation of gold in US dollars in 2021. Rodney Edvinsson, "Historical Currency Converter," Stockholm University, 2015, accessed April 16, 2021

The sculptors kept track of every gargoyle, crocket, and *fleuron* fabricated for the restoration of the cathedral.; the cost of sculptural details that can be categorized as foliate sculpture far outweighed the cost of the fabrication of gargoyles or figurative sculpture in the restoration efforts. Documents like the Duthoit invoices and Viollet-le-Duc's notes on storm damages highlight the reality of producing, maintaining, and financing a structure bristling with crockets, crowned by foliate finials, articulated by a wide range of foliate friezes. Despite Viollet-le-Duc's theories of structural rationalism, a significant portion of his restoration work focused on recreating and replacing fragile "decorative" sculpture that did not play a key role in the structural system. To Viollet-le-Duc, even gargoyles and certain kinds of foliate friezes had the function of evacuating rainwater collecting in the rooflines of medieval churches—but what about crockets, foliate finials, *rosaces*?

Viollet-le-Duc's legacy as an architectural theorist often overshadows his intense and sometimes even myopic study of architectural decoration. Given his engagement with foliate carving during restoration campaigns, it is perhaps not surprising that many essays in the *Dictionnaire* address a wide variety of foliate sculpture. One advantage of the *Dictionnaire*'s organization and format is that it can be read in infinite of ways, a point well illustrated by Barry Bergdoll in his analysis of Viollet-le-Duc's encyclopedic project.[54] While entries such as "Architecture" and "Style" have proven to be fruitful for discussions of historiography, Viollet-le-Duc also writes extensively on what could be categorized as architectural decoration or ornament; a number of entries could be considered small treatises on vegetal sculpture. For example, the entry *"Bandeau"* contains his analysis of foliate friezes; *"Bouton"* describes the rose-bud-like articulations that adorn the towers of Notre-Dame of Paris and many other Gothic

---

[54] Bergdoll, "Dictionnaire Raisonné," 2.

structures. Throughout the volumes of the *Dictionnaire*, entries such as *Chapiteau, Corniche,*

Clef, Crochet, and *Fleuron* presented new ways of seeing and understanding foliate forms,

lending a new vocabulary for describing and understanding these elements of Gothic

architecture. Some of these entries are short and function as simple definitions, while others,

such as *Chapiteau* and *Crochet,* reflect Viollet-le-Duc's deep engagement with sculptural,

decorative details. Many of these essays will be explored in the coming chapters.

The essay most relevant to this historiographic discussion is the entry titled *"Flore."* In

this essay, the architect turns his attention to theories of stylistic development: the progress and

decline of foliate forms in French medieval architecture and theories about the origins of French

foliate sculpture. He includes not only engravings of architectural details, but also illustrations of

native French plants, underscoring his thesis that the *génie* of French foliate sculpture was, in

large part, due to the sculptors' rejection of Antique and Romanesque models in favor of

observation of the natural world—and the local flora of France, in particular:

> When it comes to ornamentation, the [sculptors] no longer want to look at the old capitals
> and Romanesque friezes: instead, they go into the woods, into the fields; they look for the
> smallest plants in the grass; they examine their buds, their blossoms, their flowers, and
> their fruits, and with this humble foliage, they sculpt an infinite variety of ornaments with
> an impressive style, so confident in its execution that it far exceeds the best examples of
> Romanesque sculpture.[55]

Though Viollet-le-Duc discounts classical precedents in the production of Gothic foliate

sculpture, he allows that Byzantine and "oriental" forms from the East might have played a role

in its development. It was these sources from further afield, he writes, that inspired artists to

---

[55] "Quand il s'agit d'ornements, ils ne veulent plus regarder les vieux chapiteaux et les frises romanes : ils vont
dans les bois, dans les champs ; ils cherchent, sous l'herbe, les plus petites plantes ; ils examinant leurs
bourgeons, leurs boutons, leurs fleurs et leurs fruits, et les voilà qui, avec cette humble flore, composent une
variété infinite d'ornements d'une grandeur de style, d'une fermeté d'exécution qui laisse bien loin les
meilleurs exemples de la sculpture romane." Eugène-Emmanuel Viollet-le-Duc, "Flore", *Dictionnaire
raisonné de l'architecture française du XIe au XVIe siècle*, vol. 5 (Paris: Ve. A. Morel, 1874): 488 (my
translation).

begin to look at their own native species in the surrounding countryside. After sculpting these

petites plantes directly into the sacred space of the cathedral in the twelfth century, sculptors

began to choose more robust specimens as models for their sculpted foliage; the forms of the

later twelfth century and early thirteenth century began to bloom and grow, an evolution that

Viollet-le-Duc believes reflects the artists' own growing confidence in their craft:

> Soon (because we know that these artists do not stop along the way), they go from imitating nascent buds to developing flora: the stems lengthen and become thinner, the leaves open, spread out; the buds become flowers and fruits. Later, these artists forget their humble, primitive models: they move on to examples in the shrubs: they grasp the ivy, vines, holly, mallows, rose hips, the maple trees.[56]

Viollet-le-Duc asserts that after French foliate sculpture reached the height of sophistication and

expression in the thirteenth century, the form entered a phase of decline and decadence.

Including several paragraphs of analysis on the fleur de lys and its transformation into a

Christian and royal symbol rooted in local pagan traditions, he then stakes his claim for foliate

sculpture as a uniquely French phenomenon.

After the publication of Viollet-le-Duc's analysis, a number of scholars cultivated an

interest in medieval foliate sculpture—particularly in France. A number of subsequent studies in

the late nineteenth and early twentieth centuries engaged with botanical classification of

individual plant species of sculpted plants, formalism, and the question of medieval artists'

observation of nature. Viollet-le-Duc's strategy of illustrating plant species in concert with the

stylistic analysis of foliate sculpture took hold; the idea that foliate decoration, due to the

growing sensitivity of medieval artists to their natural surroundings, achieved full bloom in the

---

[56] "Bientôt (car nous savons que ces artistes ne s'arrêtent pas en chemin) de l'imitation de la flore naissante ils passent à l'imitation de la flore qui se développe: les tiges s'allongent et s'amaigrissent ; les feuilles s'ouvrent, s'étalent ; les boutons deviennent des fleurs et des fruits. Plus tard, ces artistes oublient leurs humbles modèles primitifs : ils vont chercher leurs exemples sur les arbustes ; ils s'emparent du lierre, de la vigne, du houx, des mauves, de l'églantier, de l'erable." Viollet-le-Duc, "Flore," 489 (my translation).

thirteenth century provided a kind of intellectual scaffolding for several scholars of French

vegetal sculpture, among them Emile Lambin and Denise Jalabert.[57]

Jalabert's contributions are particularly immense; her careful analyses of vegetal forms in

French medieval architecture include incisive, attentive descriptions that, at the time she was

writing, subtly challenged Viollet-le-Duc's naturalism thesis.[58] Although she categorized foliate

sculpture into three distinct types to show formal progression, Jalabert also recognized that

projects of taxonomy and identification were actually much more complicated within individual

monuments, and that this complexity was not easily explained by building chronology or stylistic

development. She paired observation of foliate sculpture *in situ* with stylistic analysis, noting

that sculptors employed infinite interpretations in their designs of sculpted foliage, with only

select examples referencing a recognizable plant species. Jalabert categorized this other type of

unidentifiable or hybrid foliage, often more abstracted in its forms, as *"flore généralisée."*

Importantly, Jalabert established that the naturalistic sculpted specimens coexisted with *flore*

généralisée in monuments with significant foliate sculpture. This fact had been difficult for art

historians to rationalize; observing this phenomenon at Amiens, Georges Durand posited that the

two modes of foliate sculpture were produced by two separate groups of sculptors, those who

worked on more architectural, large-scale sculpted decoration and those who took charge of the

narrative or figurative sculpture—even though there are examples of naturalistic foliage in

both.[59]

---

[57] See Emile Lambin, *La flore gothique* (Paris: André, Daly fils, 1893), *La flore de la cathédrale de Soissons et de Saint-Jean-des-Vignes* (Soissons, 1895), *Les églises des environs de Paris étudiées au point de vue de la flore ornementale* (Paris: C. Schmid, 1896); as well as Denise Jalabert, *La flore sculptée des monuments du moyen âge en France, recherches sur les origines de l'art français* (Paris: A. et J. Picard et Cie, 1965), which followed a number of articles Jalabert published on individual monuments in the 1930s.

[58] Denise Jalabert, "La première flore gothique aux chapiteaux de Notre-Dame de Paris," *Gazette des Beaux Arts*, May 1931, 283–304. Jalabert identifies a number of plants in the capitals of Notre-Dame of Paris.

[59] Durand, Georges. *Monographie de l'église Notre-Dame, cathédrale d'Amiens.* vol. 1 (of 2) Paris: A. Picard et fils, 1901.

There was a tendency in some of the earlier art historical narratives to characterize naturalistic forms, the ones with clear referents to identifiable plant species, as more advanced than examples of *"flore generalisée"* or other, abstract vegetal imagery. Nikolaus Pevsner's essay *The Leaves of Southwell* drew attention to the crisp vegetal sculpture of Southwell's compact chapter house, its naturalistic imitation of actual plants challenging "the conventions of Early English—or Early Gothic—leaf decoration," drawing inspiration "from the English countryside," a turn of phrase that echoes Viollet-le-Duc's analysis, ultimately framing this instance of foliate carving as one of the most important intellectual achievements of English medieval art, the "discovery" of observation, a subject that will be explored in the next chapter.[60] The emphasis on "naturalistic" specimens and "intellectual achievement" should give us pause to reconsider the teleological framework that has been applied to Gothic foliage; Givens' analysis of Pevsner provided a much-needed corrective that, the beauty of the Southwell carving notwithstanding, "images do not naturally evolve towards greater specificity over time."[61] Throughout the twelfth and thirteenth centuries, monuments with significant foliate sculpture incorporated a variety of modes of vegetal carving; focusing exclusively on examples of foliate sculpture that appear "naturalistic" disregards a large portion of this diverse body of work. Furthermore, decentering naturalism in medieval Europe has the potential to expand the conversation around nature in art and architectural decoration, provoking questions about the use and transmission of form, naturalism and abstraction, as well as how taxonomy and characterization play a role in shaping our understanding of works of art.

In looking to understand the historiography, many inquiries into the relationship between Gothic architecture and nature begin with the shifting developments of romanticism throughout

---

[60] Nikolaus Pevsner, *The Leaves of Southwell* (London and New York: Penguin Books limited, 1945), 11.
[61] Givens, *Observations and Image-Making in Gothic Art*, 135.

the long nineteenth century, beginning as early as the 1770s. Foliate imagery and medieval

culture became touchstones for thinkers informed by certain strains of romanticism; Germany,

England, France, and the Americas had distinct romantic literary traditions that shaped

discourses on nature, aesthetics, and art.[62] The impact of romantic writers and the development

of the environmental imagination in literature has been noted by literary scholars, while

historians of medieval architecture have highlighted the occasions that these discourses collided

with the arboreal look of Gothic architectural design: in Germany, Gothic architecture appears in

the writings of Johann Wolfgang von Goethe and Friedrich von Schlegel, who both remark on

the sylvan aesthetic of the medieval Gothic cathedral; in France, François-Réné de

Chateaubriand compared the interior of a medieval church to a cool forest; and in England,

James Hall delighted in Gothic foliate sculpture seen during his travels through France,

theorizing about their natural origins.[63]

In addition, Ruskin was especially attendant to the aesthetic effects of architectural

decoration in Gothic architecture, cultivating a spiritual view of nature informed by biblical

texts. In addition to writing about appropriate foliate ornament in the *Seven Lamps of*

---

[62] See, for example: Jonathan Bate, *Romantic Ecology: Wordsworth and the Environmental Tradition* (London ; New York: Routledge, 1991) and *The Song of the Earth* (London: Picador, 2000); Onno Oerlemans, *Romanticism and the Materiality of Nature* (Toronto: University of Toronto Press, 2004); Alain Couprie, *La nature : Rousseau et les romantiques : Chateaubriand, Lamartine, Vigny, Musset, Nerval, Hugo : textes et questions d'ensemble* (Paris: Hatier, 1985); Alison Stone, *Nature, Ethics, and Gender in German Romanticism and Idealism* (London ; Lanham, MD: Rowman & Littlefield Publishers, 2018); Kenneth Daley, *The Rescue of Romanticism: Walter Pater and John Ruskin* (Athens: Ohio University Press, 2001). Literary scholars have identified and analyzed the underlying nationalist projects inherent in romanticist ideas about art and nature. See David Aram Kaiser, *Romanticism, Aesthetics, and Nationalism*, Cambridge Studies in Romanticism 34 (Cambridge, U.K.; New York: Cambridge University Press, 1999).

[63] Johann Wolfgang von Goethe, "Von Deutscher Baukunst," in *Von Deutscher Art und Kunst*, ed. Heinz Kindermann, Deutsche Literatur; Sammlung Literarischer Kunst- Und Kulturdenkmäler in Entwicklungsreihen. Reihe 15: Irrationalismus (Leipzig: P. Reclam jun, 1935 [original printed 1773]), 209–36; Friedrich von Schlegel, "Principles of Gothic Architecture," in *The Aesthetic and Miscellaneous Works of Frederick von Schlegel*, trans. E. J. Millington (London: G. Bell and Sons, 1875 [German printed 1804/1805), 149–199; François-René Chateaubriand, *Génie du Christianisme: ou beauté de la religion chrétienne*, 5 vols. (Paris: Chez Migneret, imprimeur, 1802), vol. 1, 184; James Hall, *Essay on the Origin and Principles of Gothic Architecture* (Edinburgh, 1797).

Architecture and *The Stones of Venice*, Ruskin rhapsodized about the foliate sculpture at Notre-Dame of Amiens specifically, both in the cathedral's portals and the carving of the sixteenth-century choir stalls.[64] Pevsner set Ruskin's romantic view of the medieval cathedral against the structural rationality of Eugène-Emmanuel Viollet-le-Duc—and indeed, these writers disagreed in their views on restoration and medieval aesthetics.[65] But when it came to imagining medieval sculptors and their choice to depict the natural world, both Ruskin and Viollet-le-Duc saw more than sculptures of plants and representations of individual specimens: they saw also evidence of medieval artisans engaging deeply with the natural world and rejoicing in its plant life, not unlike their nineteenth-century contemporaries in the literary sphere.

Questions about naturalism and sculpted plant life have persisted in art historical scholarship of the last several decades. While Givens' interventions are nuanced and rich is historiographic detail, covering a variety of media, some art historians have focused on the treatment of plant life in sacred space. Sculpted plants are often omitted from narratives addressing Gothic sculpture, though some scholars see the treatment of plant life as part of growing tendency to represent figurative and vegetal subjects naturalistically.[66] Other art historians view figurative sculpture and vegetal sculpture as having two very different artistic approaches. Michael Camille, for example, posited that medieval sculptors could apply observation and naturalism to plant life but not figurative sculpture, because foliate forms were not loaded with dangerous meaning:

---

[64] John Ruskin, *Our Fathers Have Told Us: The Bible of Amiens, Chapter IV, Interpretations*, 4th ed. (London: George Allen & Sons, 1909).

[65] Nikolaus Pevsner, *Ruskin and Viollet-Le-Duc: Englishness and Frenchness in the Appreciation of Gothic Architecture*, Walter Neurath Memorial Lectures 1969 (London: Thames & Hudson, 1969); see also David Spurr, "Figures of Ruin and Restoration: Ruskin and Viollet-Le-Duc," in *Architecture and Modern Literature* (University of Michigan Press, 2012), 142–61.

[66] For example, Max Dvořák, *Idealism and Naturalism in Gothic Art* (Notre Dame, Ind.: University of Notre Dame Press, 1967), 32; Frankl, ed. Crossley, *Gothic Architecture*, 293; Willibald Sauerländer, *Gothic Sculpture in France 1140-1270* (London: Thames and Hudson, 1972), 42-43.

Rather there were different degrees of "natural" that a sculptor or painter could deploy, applicable to different categories of persons and things. So, while a stone flower might have been carved by observing an actual flower, the human body was structure out of schematic shapes of the divine geometer. The fallen human body was far too loaded with taboos and dangers to be portrayed with the verisimilitude of a leaf. Adam and Eve, the last and most God-like of things created, were also the last to be naturalistically represented in art.[67]

It is worthwhile to consider these theories in concert with visual evidence, which suggests that artists did not treat vegetal subjects any differently than their human counterparts. The thirteenth-century freestanding sculpture of Adam from Notre-Dame of Paris, now conserved in the Musée de Cluny, shows an attentive naturalism in depicting the flesh of the human body (fig. 1.1). In Villard de Honnecourt's *cahier*, a schematized Madonna and Child shares the same space as two sprouting leaves, one that is clearly modeled on a five-point star and the other, which could be either a sprouting leaf or a diagram of a sexfoil (indeed, this foliate diagram quickly resembles the ubiquitous foliate designs found in contemporary architecture, quatrefoils and sexfoils).[68] Both religious subject and foliage are depicted as possessing an underlying geometric truth (fig. 1.2). These diagrams follow a number of schematized drawings in Villard's *cahier* that depict human figures, human faces, as well as architecture that is composed of an underlying geometry (fig. 1.3, 1.4). Seeing the geometry in all things plant, animal, and human was part of a visual representational system—this approach, however, is not consistent for all of Villard's drawings. As his drawings of the male nude indicate, the artist was sometimes interested in reaching for other systems of representation, copying, and image-making (fig. 1.5, 1.6). And while many of

---

[67] Michael Camille, *Gothic Art: Glorious Visions* (New York: Harry N. Abrams, 1996), 134.

[68] For a variety of approaches to the cahier of Villard de Honnecourt, see Carl F. Barnes, *Villard de Honnecourt--the Artist and His Drawings: A Critical Bibliography*, A Reference Publication in Art History (Boston, Mass: G.K. Hall, 1982); Roland Bechmann, *Villard de Honnecourt: La pensée technique au XIIIe siècle et sa communication* (Paris: Picard, 1991); Stephen Murray, "Villard de Honnecourt: Ymagier and Interlocutor," in *Plotting Gothic* (Chicago: University of Chicago Press, 2014), 17–46, Jean Wirth, *Villard de Honnecourt: architecte du XIIIe siècle*, Titre Courant 58 (Geneva, Switzerland: Librairie Droz S.A., 2015).

his images are a fantasy or approximation of reality, Villard was clearly deeply engaged with plants, animals, technology, and architecture—and how these subjects converged and overlapped (fig. 1.7, 1.8). The images in the *cahier* alone challenge the idea that naturalism was saved for plants simply because they were a less controversial subject. The theory of interpretive neutrality or "safety" of plant imagery in Christian art should be regarded with caution; as we will see in the next section, the representation of plants could be quite potent in religious signification. The lack of stability of these foliate image systems, however, made it difficult for art historians to decode these systems with the same confidence as other types of medieval Christian art.

1.3 Finding Meaning in Sculpted Plants: Problems of Iconography and Ornament

Whether identifiable species or not, many examples of sculpted foliage do not appear to have a stable meaning, nor can they always point directly to an iconographic tradition—and upon closer examination, many subjects depicted in medieval art share these characteristics. Yet, within the context of medieval Christianity, the role of plant symbolism in biblical texts could make vegetal imagery potent and evocative. Sculptors, glaziers, and manuscript illuminators depicted many of these plant-laden images and metaphors—the Living Wood of the Cross, the Tree of Jesse, the Tree of Knowledge in Paradise, Paradise itself—developing their own iconographical traditions transmitted across time and media. The artists producing these images did not need to draw upon their knowledge of botanical specimens or heightened realism in order to contribute to or perpetuate these image systems. Medieval depictions of the Tree of Jesse, for example, often employed intertwining vines, leaves, and symmetrical "branches" to signify a tree-like structure; the species of the tree and its botanical accuracy was less important than the metaphor of progeny that the vegetal imagery represented. Within the context of sacred space, some examples of sculpted foliage may suggest theological meaning, but it matters greatly how this iconography is being exercised and how the specific, local context of the monument can be taken into consideration.

Applying iconographic frameworks to sculpted plants has been a mixed enterprise. Looking for theological and philosophical meaning in the sculpted plants of medieval France, Émile Mâle adopted the stylistic development theories of Viollet-le-Duc but tried to imbue vegetal forms with the symbolism present in the texts of medieval writers and several types of medieval art. Referencing Vincent of Beauvais, Mâle argues that natural phenomena were interpreted symbolically, the visible acting as a window to suggest the invisible. At the center of these representations of nature, he argues, is love and passion on the part of the artists.

> But we only have to raise our eyes to see the grapevine, the raspberry bush laden with fruit, and the long shoots of the wild rose clinging to the archivolts. ... How little do the old masters deserve the charge of incompetence and sterility. Never have artists been more impassioned by the beauties of nature. Their cathedrals are movement and life itself.[69]

In extolling the beauties of foliate sculpture, we might see similarities with the approaches of Ruskin and Viollet-le-Duc; Mâle sees this work as proof of the underlying genius of the medieval artist. Highlighting the skill and artistry of these artisans is a particular focus in his study, but he also looks to medieval theology and philosophy to show his readers what could be gained by applying textual analysis to the interpretation of foliate sculpture. His examples range from Peter of Mora, meditating on roses in his garden ("the rose is the choir of martyrs, and again, it is the choir of virgins. When [it] is red, it is the blood of those who died for the faith...") to Adam of St. Victor's likening of a walnut to the flesh of Jesus Christ, his divinity, and the

---

[69] Emile Mâle, *Religious Art in France, the Twelfth Century: A Study of the Origins of Medieval Iconography*, Bollingen Series, 90:1 (Princeton, N.J: Princeton University Press, 1978), 31. Originally published in 1898: Emile Mâle, *L'art religieux du XIIIe siècle en France: étude sur l'iconographie du Moyen Age et sur ses sources d'inspiration* (Paris: Ernest Leroux, 1898).

wood of the cross. Following Viollet-le-Duc's analysis, Mâle readily adopts the idea that the thirteenth century represented a new era for the representation of nature in art. He repackages the now-familiar argument of stylistic development of sculpted foliage, drawing a distinction between Romanesque forms, which he says are copied from antiquity or textile designs from the East, and thirteenth-century forms, which take their inspiration from the local flora and fauna of France—another echo of Viollet-le-Duc's *petites plantes* theory.

Mâle holds back, however, in applying any theories of style or text-based interpretation to individual monuments. It is clear from his joyful descriptions that he has looked pleasurably upon the diversity of foliate sculpture from a variety of medieval French buildings, writing about the general beauty and ingenuity of foliate sculpture. But at one point in his analysis, he asks if there is anything to be gained by probing the symbolism of sculpted plants, even as he has shown that medieval thinkers were interested in finding the divine in observable, natural phenomena— and even in individual specimens, as the moralistic herbaria and bestiaries of the period attest. No, he writes, the "flora and fauna of the Middle Ages, whether real or fantastic, have for the most part *only a decorative function*."[70] Texts, he concludes, cannot be used to decipher or explain the presence of sculpted foliage in the cathedrals of Reims, Paris, Amiens, Sens, Laon, or Rouen. We can only understand the artists' enthusiasm for nature through these forms—nothing about foliate sculpture in sacred space, he believes, can tell lend us further understanding of religious belief or the practice of Christianity of the Middle Ages.

Perhaps these ideas were informed by French medieval sources that comment on the symbolism of the church, its materiality, and decoration—but are surprisingly sparse on the subject of foliate sculpture. Abbot Suger, for example, details the rich materials and architectural

---

[70] Mâle, *Religious Art*, 51 (my emphasis).

design of his chevet at Saint-Denis, but only writes in passing about the foliate sculpture of the capitals and does not offer any theories as to their meaning. Similarly, William Durandus analyzes many parts of the sacred space of the cathedral but does not mention the symbolism of foliate sculpture. In English sources, Gervase of Canterbury comments on the "excellent carving" of the new capitals at Canterbury but does not offer any explanation for the changes in the foliate sculpture from the old work to the new. In the Metrical Life of St. Hugh, the author writes a veritable sermon on the symbolism of the various stones used in the construction of the rebuilt cathedral, with many prescient observations between the relationship between nature and architecture, but the sculpted foliage—ever prominent at Lincoln Cathedral—is not specifically mentioned. If medieval chroniclers failed to notice foliate sculpture, why address this art form through the lens of today's art historical inquiries?

The general scarcity of medieval written sources on foliate sculpture in architecture seem to be at odds with the physical presence of this sculpted decoration, which has captured the attention of so many art historians—even those that dismiss it as an additive decorative treatment lacking interpretive depth. With a dearth of medieval texts that address sculpted foliage directly, scholars like Mâle appear to even dismiss their own observations. Notably, however, many examples of foliate sculpture—in individualized capitals in a dado, fields of foliate relief, foliate frames above benches intended for seated canons in chapter houses—are often placed at eye-level, as if to engage the viewer in a conversation about the natural world in the context of sacred space. The recurrence of this phenomenon across building sites and geographical areas has yielded some thoughtful analysis, though art historians' interpretations of foliate sculpture has often hinged on medieval texts exploring Platonism, Neo-platonism, and natural philosophy (an approach that will be explored in more detail in the next chapter).

As noted above, however, select species of plants have a rather stable symbolic meaning

in medieval Christianity. The vine, for example, references Christ, while the white rose came to

accompany and sometimes even stand in for Marian imagery. References to fecundity, plant life,

blossoming trees and metaphor grounded in the natural world appear in great numbers in biblical

texts; the natural world in the Bible is both material and medium: it is created by God and

ultimately used to communicate messages to mankind through withering plants, natural disasters,

and fertile fields.[71] Natural themes abound in the decoration of Christian sacred space, and as

extant mosaics from Santa Constanza in Rome as well as San Vitale and Sant'Apollinare Nuovo

in Ravenna attest, the use of plant life in sacred architecture was already a convention in church

decoration in the Christian world before patrons and master masons were experimenting with

Gothic forms.[72] What makes the foliate decoration in Gothic architecture difficult to quantify,

however, is the full integration of foliate forms into the architectural elements themselves: rose

windows, trefoils and quatrefoils, tracery, crockets. These elements were not only a decorative

overlay atop the underlying architecture; in many cases, foliate forms became the architecture, a

merging of structure and ornament that eluded some twentieth-century discourses on the subject.

---

[71] A. W. Anderson, *Plants of the Bible* (London: Lockwood, 1956); Harold N. Moldenke and Alma L. Moldenke, *Plants of the Bible*, A New Series of Plant Science Books, v. 28 (Waltham, Mass: Chronica Botanica Co, 1952); Michael Zohary, *Plants of the Bible: A Complete Handbook* (London ; New York: Cambridge University Press, 1982). In addition to the *lignum vitae*, the Tree of Life, in Genesis 2:9 and Revelation 21:1-2, the Psalms and especially the Canticles call on several distinct kinds of plants, flowers, and trees. Plant life also plays a narrative function in many biblical stories, such as the story of Moses and the burning bush and the blossoming rod of Aaron. In many instances in the texts, God communicates messages directly through the medium of blossoming or withering plants. Interest in the specific plants of the geographic regions mentioned in the Bible have generated reference tools of biblical botany; Harold Moldenke and A. W. Anderson catalogued biblical plant references, matching the text to an illustration of a botanical specimen and a description of each plant mentioned; these efforts have been redoubled by Michael Zohary and others. The symbolic, non-botanical dimension of plant life in biblical stories, on the other hand, has been examined almost exclusively by art historians.

[72] Finbarr Barry Flood, *The Great Mosque of Damascus : Studies on the Makings of an Ummayad Visual Culture* (Boston: Brill, 2001), 87-88. Mosaics that feature natural forms and greenery in architectural settings are prevalent in Byzantine and Islamic architecture, but perhaps the most commented-upon example is the vine mosaic that articulates the interior elevation of the Dome of the Rock

Though these churches were often painted, plastered, gilded, and adorned with precious textiles and metalwork, this integration suggests a reimagining of sculptural and architectural form that eluded art historical methods based on clear delineations of media and form.

Beyond the classic definitions of iconography put forth by Warburg, Wittkower, Panofsky, and others, Gombrich and Krautheimer demonstrated the importance of accounting for elusive meanings and multivalence in art historical interpretation.[73] A notable group of German art historians applied iconographic themes and theological meaning to sculpted foliage within architectural space, limiting their focus to late Gothic examples, but the iconographic analysis of sculpted foliage seems to have been discounted along with harsh critiques laid against Sedlmayr's characterization of the Gothic cathedral as the actual representation of Heaven.[74] Iconographic interpretations of foliate forms, however, need not be eliminated from the discourse, especially if the interpretation acknowledges the complexities of vegetal forms and architectural decoration.

Finally, the characterization of foliate sculpture as 'ornament' and the rise of ornament studies in non-Western art has further complicated these narratives. Riegl's characterization of "ornament" as having its origins in natural forms helped to define foliate decoration as separate from narrative sculpture and architectural space, a distinction that pushed art historians to

---

[73] E. H. Gombrich, "Aims and Limits of Iconology," in *Symbolic Images* (London: Phaidon, 1972); Erwin Panofsky, "Iconography and Iconology: An Introduction to the Study of Renaissance Art," in *Meaning in the Visual Arts: Paper in and on Art History* (Garden City, NY: Doubleday, 1955), 26–54.; Richard Krautheimer, "Introduction to an 'Iconography of Mediaeval Architecture,'" *Journal of the Warburg and Courtauld Institutes* 5 (1942): 1–33.

[74] Karl Oettinger, "Laube, Garten und Wald. Zu einer Theorie der Stüddeutschen Sakralkunst 1470-1520," in *Festschrift für Hans Sedlmayr*. München, 1962: pp. 201-228, M. Braun-Reichenbacher, *Das Astund Laubwerk, Entwicklung, Merkmale und Bedeutung einer spätsgotischen Ornamentform*. Nürmberg, 1966; F. Bösch-Supan, *Garten und Landschafts und Parndiesmotive im Immenraum, Eine ikonographische Untersuchung*. Berlin, 1967. On Sedlmayr in context, see Crossley, Paul. "Medieval Architecture and Meaning: The Limits of Iconography." *The Burlington Magazine* 130, no. 1019 (1988): 116–21.

broaden classifications of genre and form.[75] Henri Focillon considered ornament to be a frontier,

recognizing that even simple flourishes shape the very framework, aesthetic experience, and

form of a work of art:

> Even before it becomes formal rhythm and combination, the simplest ornamental theme, such as a curve or *rinceau* whose flexions betoken all manner of future symmetries, alternating movements, divisions, and returns, has already given accent to the void in which it occurs, and has conferred upon it a new and original existence. Even if reduced merely to a slender and sinuous line, it is already a frontier, a highway. Ornament shapes, straightens, and stabilizes the bare and arid field on which it is inscribed. Not only does it exist in and of itself, but it also shapes its own environment.[76]

This is especially evident in Gothic architecture, where the merging of ornament and structure

occur. Without applying this theory to specific monuments or works of art, Foçillon recognized

the value of considering decorative details integral to aesthetic experience and formal impact. "If

we follow the metamorphoses of this form," he continues, "if we study not merely its axes and

its armature, but everything else that it may include within its own particular framework, we will

then see before us an entire universe that is partitioned off into an infinite variety of blocks of

space."

Restoring the three-dimensionality and architectonic qualities of foliate forms might add

another rich interpretive dimension to the study of architectural decoration. Margaret Graves

recently highlighted the interpretive difficulties of applying theories of ornament studies built on

the foundations of Reigl's theories and the flattening decontextualization of ornamental motifs in

Owen Jones' albums from *The Grammar of Ornament*, stripping plastic arts from their

---

[75] Alois Riegl, *Stilfragen: Grundlegungen zu einer Geschichte der Ornamentik*, 2. aufl (Berlin: R. C. Schmidt & co, 1923). English translation: *Problems of Style: Foundations for a History of Ornament*, trans, David Castriota, (Princeton, NJ: Princeton University Press, 1992). For Riegl in historiographic context, see Arturo Carlo Quintavalle, "The Philosophical Context of Riegl's 'Stilfragen,'" *Architectural Design* 51, no. 6 (January 1, 1981): 17.

[76] Henri Focillon, *The Life of Forms in Art*, ed. Charles Beecher Hogan and George Kubler, Yale Historical Publications; History of Art, IV (New Haven; London: Yale University Press; H. Milford, Oxford University Press, 1942), 20.

architectural contexts.[77] Recognizing the specific challenges for architectural decoration, Graves

writes,

> …ornament is recognized at best as a cognitive system overlaying a form, at worst as an
> arbitrary or superficial afterthought that could be skimmed off one surface and stuck onto
> another without disruption to its inherent nature. … Architecture *as* ornament was not
> normally recognized as a distinct category in spite of the relatively common occurrence
> of certain forms, particularly arches and arcades. This exclusion is symptomatic of a
> recurring uncertainty about how to treat architectural motifs within any categorizations of
> theories of ornament.[78]

Graves' insights on this "zone of ambiguity" that ornament, including foliate decoration,

occupies, is important in re-framing foliate sculpture in Gothic architecture—and, arguably, any

art form that engages the architectural space as a medium. Graves argues that architectural media

are, in fact, too complicated to fit realistically within the narrow discourse of ornament studies,

which does not account for their three-dimensionality in architectural contexts. In a similar vein,

foliate forms in the well-studied, canonical monuments of medieval France and England cannot

be convincingly accommodated by existing discourses on architectural sculpture, architectural

form, and interpretation based on iconography, extensive as it is in the scholarship of Western

medieval art. Nor can foliate sculpture be wholly incorporated into theories of aniconic imagery

or abstraction, as it sometimes references actual plants—often at eye-level of the viewer—or is

made to look "plant-like" for aesthetic effect.

1.4 Conclusion: Sculpted Foliage and the Medieval Environmental Imagination

Many opportunities lie in studying what this sculpture does for sacred architecture, why it

was incorporated so thoroughly into some monuments but not in others, as well as identifying

---

[77] Owen Jones, *Grammar of Ornament* (London: Bernard Quaritch, 1868).
[78] Margaret S. Graves, *Arts of Allusion: Object, Ornament, and Architecture in Medieval Islam* (New York,
NY: Oxford University Press, 2018), 63.

specific cultural-architectural contexts to untangle how foliate imagery is at play in larger,

overarching iconographic programs. The project of seeing this sculpture anew beyond the

structure-ornament dichotomy that governed much of twentieth-century architectural history is

still very much underway. Given the nature of Gothic systems of architecture and decoration,

intertwined as they are, these monuments offer many opportunities to imagine how we might

undertake a study of architecture and decoration beyond the structure-ornament binary, as

theories of artistic integration and studies of "transitional" monuments have been able to

accomplish.[79] As these examples have shown, the study of Gothic architecture is also deeply

enmeshed with the status of its vegetal ornament. In attending to examples of foliate sculpture,

we may be able to reclaim and rediscover this architectural system anew.

But to begin to untangle this web of signification, a return to the medieval written sources

might help illuminate the sources of medieval sculptors' deep engagement with plant life. One

common thesis in the study of naturalistic plant sculpture was that the influx of Aristotelian

ideas, spread through medieval universities and cathedral schools, somehow prompted sculptors

to observe from nature, rather than copy from templates or use abstract forms. Another area of

opportunity for sculpted foliage lies in object-based study; as recent scholarship on medieval art

and nature is anchored in medieval texts and historiography, looking closely at the objects

themselves, especially those that are impossible to identify from a botanical perspective, may

yield more information about their manufacture, use, and importance. Finally, while ecocritical

theory has been embraced by scholars of medieval history and literature, environmental

entanglements within this framework remain relatively unexplored in the analysis of medieval

---

[79] Virginia Chieffo Raguin, Kathryn Brush, and Peter Draper, eds., "Artistic Integration in Gothic Buildings," electronic resource, *JSTOR EBooks*, 1995 Sankovitch, Anne-Marie. "Structure/Ornament and the Modern Figuration of Architecture." *The Art Bulletin* 80, no. 4 (1998): 687–717.

architecture and its foliate sculpture, though this is quickly changing.[80] Historical climatology,

which provides a number of datasets on climactic change throughout the Middle Ages in

increasing detail and specificity, also represents an opportunity for studies of the medieval

natural world and its figuration in text and images. Finally, attending to foliate sculpture gives art

historians another way study what we might call the medieval environmental imagination, a

collaboration between the artisans who specialized in this craft, the masons who envisioned the

buildings' design, and the clerical authorities who, in some instances, seem to have viewed

monumental foliate sculpture as a powerful commentary on spiritual power. The following

chapters will explore these themes by re-evaluating the place of foliate sculpture in the material-

historical record of France and England, while providing a framework for future studies of non-

figurative foliate sculpture and architectural decoration.

---

[80] See, for example, Anne F. Harris, "Water and Wood: Ecomateriality and Sacred Objects at the Chapel of
Saint-Fiacre, La Faouët (Brittany)," *Journal of Medieval and Early Modern Studies*, 2014, 585–615 and
"Hewn," in *Inhuman Nature*, ed. Jeffrey Jerome Cohen (Washington, D.C.: Oliphaunt Books, 2014), 17–38;
Danielle B. Joyner, "A Savin Bush in the Cloister: Art and Nature in the Plan of St. Gall," *Studies in
Iconography* 38 (2017): 55–106. Before the ecocritical turn, see Roland Bechmann, *Les racines des
cathédrales : l'architecture gothique, expression des conditions du milieu* (Paris: Payot, 1981) and *Des arbres
et des hommes : la forêt au moyen âge* (Paris: Flammarion, 1984), translated into English in 1990; and
medieval history in environmental studies, especially Lynn White, "The Historical Roots of Our Ecologic
Crisis," *Science* 155, no. 3767 (1967): 1203–7; Bron Taylor, Gretel Van Wieren, and Bernard Daley Zaleha,
"Lynn White Jr. and the greening-of-religion hypothesis," *Conservation Biology* 30, no. 5 (2016): 1000–1009;
Anna Peterson, ed., *Religion and Ecological Crisis: The "Lynn White Thesis" at Fifty* (Routledge, 2016).

2

Who Discovered Nature?

Philosophy and Practice in England and France

Untangling the threads of philosophical thought in medieval intellectual circles is nearly as challenging as cataloguing the great diversity of sculpted plants in the jambs of cathedrals in France and England. This chapter will explore treatises on natural philosophy and practical guides to plant cultivation in the medieval West, examining how surviving written sources on the natural world might aid and intersect with the study of foliate sculpture. Historiographic questions about the utility of intellectual history are central to understanding how foliate sculpture has been studied in the recent past and how to proceed with inquiries into its interpretative potential. In reviewing trends in medieval natural philosophy as well as written work on plants, weather, and agricultural cultivation, it becomes clear that attitudes towards nature shifted in intellectual circles over the course of the twelfth and thirteenth centuries. It is possible, moreover, that favorable growing seasons and a climactic change contributed to interest in meteorological phenomena. The specific interest in plants reaches an apex in the late thirteenth century writings of Albertus Magnus, whose theories on leaf formation and morphology echo theories of the humors in the human body, describing the plant body as a living organism subject to universal laws governing growth and regeneration. The examination of a timeline of intellectual activity allows us to modify our understanding of foliate sculpture, which appears in great variety already in the latter half of the twelfth century. While many questions remain, it might be appropriate to see these sculptures as a milieu through which medieval

49

thinkers could conceptualize nature within the built environment, rather than evidence that

patrons, masons, and sculptors were engaging deeply with natural philosophy.

2.1 Intellectual History and Medieval Art History

The twelfth century was a period of tumultuous change in Western medieval culture. The

explosion of monastic reform movements, growing popularity of the cult of saints and

pilgrimage, organization of cathedral schools, and influx of newly translated texts reordered the

Church hierarchy, shifted power dynamics, and opened new avenues for intellectual inquiry.

Technological advancements laid the foundation for vast expansion in agriculture and industrial

production. In the city of Amiens, the local woad industry experienced significant growth during

this time, becoming a profitable center of textile production strategically located for English

wool traders, textile artisans in the Low Countries, and the markets of Paris.[81] This was also a

period of experimentation and expansion in the production of art and architecture, a formative

period in which masons and their patrons began to experiment with new architectural forms in

stone that would become widely used in the thirteenth century. The study of the relationship

between medieval art and intellectual history has a rich historiographic tradition, a well-worn

path that has been commented upon and debated at length.[82] To understand whether

---

[81] A. de Calonne, *Histoire de la ville d'Amiens*, vol 1. Bruxelles 1976 (originally published 1899), 128-9 and
200-10; Robert Fossier, *La terre et les hommes en Picardie jusqu'à la fin du XIIIe siècle*, vol. 1 (Paris,
Louvain: B. Nauwelaerts, 1968), 422-6; Murray, *Amiens*, 22; Bernard Verhille, "Le commerce médiéval des
plantes tinctoriales dans le nord de l'Europe," *Bulletin de l'Association des amis de la cathédrale d'Amiens* 25
(2010): 41–45; Roger Wadier, *L'or bleu de Picardie* (Editions Cathers Dupays, 2016).

[82] For example, the rich discussion of Gothic architecture and scholastic philosophy in the wake of the
publication of Erwin Panofsky's *Gothic Architecture and Scholasticism* lectures, which first appeared in print
in 1951. See Peter Kidson, "Panofsky, Suger and St Denis," *Journal of the Warburg and Courtauld Institutes*
50 (1987): 1–17; Charles Radding and William W. Clark, *Medieval Architecture, Medieval Learning: Builders
and Masters in the Age of Romanesque and Gothic* (New Haven, CT: Yale University Press, c1992). The
relationship between Gothic design and geometrical knowledge diffused through the intellectual circles of the
University of Paris is explored in Stefaan Van Liefferinge, "The Hemicycle of Notre-Dame of Paris: Gothic

experimentation in foliate sculpture is related to parallel intellectual developments, however, it is

useful to consider how the disciplines of intellectual history and art history have utilized material

evidence in concert with contemporary texts.

As noted above, art historians have long employed intellectual history in service of

interpreting foliate sculpture. Several studies have attempted to explain "trends" in foliate

carving and growing interest in plant life representation as an expression of the shifting

philosophical tides of the twelfth and thirteenth centuries. These studies have placed this

sculpture within the context of the intellectual interests in Platonism and transition to

Aristotelianism; as foliage became more realistic, it came to reflect ideas that are entrenched in

Aristotelian thought: forms based on observation, rather than divine geometry.[83] Perhaps this

tendency grows out of theories of Christian iconography on which many of the first art historical

studies of medieval art are based; because foliate decoration cannot always be fully explained by

a Biblical text, scholars sought additional sources that could lend a text-based interpretive

framework to works of art that depict no narrative and appear superfluous to the moralizing

iconographic programs of church portals. Indeed, there are many such sources to consider, as

many medieval scholars and theologians were quite preoccupied with change and transformation

in the natural world. Lottlisa Behling's study of sculpted foliage reads like a detailed study of

medieval philosophical thought with an extensive catalogue of photographs of sculpted foliage.[84]

Jean Givens' richly historiographic study on Gothic naturalism employs textual sources to

discuss a variety of manuscripts and examples of sculpture. These studies and others like them

---

Design and Geometrical Knowledge in the Twelfth Century," *Journal of the Society of Architectural Historians* 69, no. 4 (2010): 490–507.

[83] Otto Pächt, "Early Italian Nature Studies and the Early Calendar Landscape," *Journal of the Warburg and Courtauld Institutes* 13, no. 1/2 (1950): 13–47. Pächt's essay on botanical illustration and medieval calendars, which is often a starting point and/or flashpoint for art historians studying medieval depictions of nature.

[84] Lottlisa Behling, *Die Pflanzenwelt der Mittelalterlichen Kathedralen* (Köln: Böhlau Verlag, 1964).

suggest a proto-Renaissance framework in which artists who "discovered" the observation of nature were somehow connected to those studying and teaching Aristotelian ideas in universities.[85] They consider foliate sculpture to be explained, at least in part, by developments in intellectual history.

Scholars of intellectual history, on the other hand, largely dismiss foliate sculpture as a medium through which medieval attitudes towards nature can be understood. Marie-Dominique Chenu's foundational study on the renaissance of the twelfth century describes a decisive paradigm shift, an awakening to the possibilities of observation, and the dissemination of the study of cause and effect across intellectual disciplines: law, literature, religion, and art. Chenu states in his introduction that he is not concerned with foliate sculpture, even though his study addresses transformation in the realm of art production quite directly, including remarks on sculpted plants.[86] More recently, Jeffrey Jerome Cohen outlined a similar occlusion to sidestep architectural sculpture in his volume on the materiality of stone, citing the fact that "lithic sculpture tends towards the anthropomorphic" and is "as plastic a substance as wood" in the

____

[85] It is important to note that Aristotelian texts had been studied for centuries in Arab and Byzantine circles prior to their translations into Latin. To claim 'discovery' is erroneous and assumes the knowledge had been "lost." Rather, medieval Western European scholars operated within a linguistically sequestered intellectual *habitus* that was in the process of transformation. See Lawrence P. Schrenk, ed., *Aristotle in Late Antiquity*, Studies in Philosophy and the History of Philosophy, v. 27 (Washington, D.C: Catholic University of America Press, 1994); Ahmed Alwishah and Josh Hayes, eds., *Aristotle and the Arabic Tradition* (United Kingdom: Cambridge University Press, 2015); Michele Trizio, "Reading and Commenting on Aristotle," in *The Cambridge Intellectual History of Byzantium*, ed. Anthony Kaldellis and Niketas Siniossoglou (Cambridge: Cambridge University Press, 2017), 397–412.

[86] "The discovery of nature: we are not now concerned merely with the feeling for nature which the poets of the time evinced here and there in fashionable allegorical constructions, nor are we concerned merely with the plastic representations of nature that sculptors fashioned at the portals and on the capitals of cathedrals." Marie-Dominique Chenu, *Nature, Man, and Society in the Twelfth Century; Essays on New Theological Perspectives in the Latin West*, trans. Jerome Taylor and Lester K. Little (Chicago: University of Chicago Press, 1968), 4. Chenu goes on to quote Lewis Mumford, who included sculptors' foliate sculpture in his analysis of the new, thirteenth-century sensibility for naturalistic art and technological improvement. Chenu agrees with this assessment, claiming that the "rise of new techniques both betokened and promoted a true discovery, an active discovery of nature; and man advanced toward self-discovery as he came thus to master nature." Chenu, *Nature*, 39.

hands of an artisan.[87] What if scholars of intellectual history and medieval culture considered the vast record of foliate sculpture as thoroughly as art historians considered medieval texts on natural philosophy? Even within this narrow inquiry, the primacy of text over visual evidence has shaped the way foliate sculpture has been examined and studied. This distance could be bridged with an analysis of the medieval texts at hand in concert with a basic chronology of foliate sculpture as it appeared in northern France and England.

2.2 Natural Philosophy and the Study of Plants in the Medieval West

Latin translations of the work of Aristotle and Avicenna would transform the intellectual centers of medieval France, England, and continental Europe, but it is important to acknowledge that the hunger for understanding the natural world was already sharpened before these texts were disseminated. Scholars went to great rhetorical lengths to try to understand the workings of the natural world with the intellectual tools they had on hand. They understood plant life and the patterns of the natural world through the lens of the early Christian fathers and ancient texts kept in monastic libraries, especially Plato's *Timeaus*. The *Timaeus* emphasizes the role of the divine craftsman in imposing order on the chaotic universe, suggesting that the world itself is an ordered organism through which mathematical truths and proportions can be revealed. Timeaus' speech draws a distinction between that which always is and has no becoming and that which *becomes*, but never *is*; everything that comes to be must do so by agency of some cause.[88] This text had a major impact on medieval philosophers preoccupied with the idea of the *animus*

---

[87] Jeffrey Jerome Cohen, *Stone: An Ecology of the Inhuman* (Minneapolis: University of Minnesota Press, 2015), 13, note 38.
[88] Plato, *Timaeus*. Donald J. Zeyl, trans. (Indianapolis: Hackett Pub. Co, 2000).

mundi, or world soul, though their contact with this text was often diffused through other sources

and interpreted in a number of ways.

Most medieval scholars accessed the *Timaeus* through a commentary by Chalcidius;

others absorbed Platonic ideas from Augustine, the Gospel of St. John, or other early Christian

writers.[89] Twelfth-century writers aware of the *Timeaus* sought to reconcile Plato's creation story

with Genesis; Peter Abelard and William of Conches identified the *animus mundi* with the

Christian Holy Spirit—the former famously ran into trouble for his ideas, but the latter was able

to sidestep controversy rhetorically.[90] Deep in rhetorical analysis, writers explored the

implications of this idea, including the question of whether the existence of an *animus mundi*

meant that each living thing—humans, animals, and plants—had two souls. Bernard Silvester,

Honorius of Autun, and Alain de Lille were among the cohort of scholars exploring these ideas,

though few writers actually went to the trouble to identify the plant world as a separate

intellectual entity worthy of contemplation.

Some writers, however, did come to recognized that plant life, in a Platonic sense, is

perpetually in a state of becoming. William of Conches, for example, references plants briefly in

his *Philosophia Mundi*, even though plants themselves were never the focus of his philosophical

treatise.[91] Literature about plant life tended toward the practical: Theophrastus' *De plantis*,

thought to have been written by Aristotle, addressed the care and cultivation of plants and the

---

[89] Edward Grant, *A History of Natural Philosophy: From the Ancient World to the Nineteenth Century* (New York: Cambridge University Press, 2007); Charles Burnett, "The Introduction of Aristotle's Natural Philosophy into Great Britain : A Preliminary Survey of the Manuscript Evidence," *Rencontres de Philosophie Médiévale*, 1996, 21–50; Marie-Dominique Chenu, "The Platonisms of the Twelfth Century," in *Nature, Man, and Society in the Twelfth Century; Essays on New Theological Perspectives in the Latin West*, trans. Jerome Taylor and Lester K. Little (Chicago: University of Chicago Press, 1968), 49–98; Benedicta Ward, "The Theory of Miracles" in *Miracles and the Medieval Mind : Theory, Record and Event 1000-1215* (Aldershot: Scolar, 1987) 3-20.
[90] "According to some the soul of the world is the Holy Spirit." William of Conches, *Philosophia mundi*, ch. 4, ed. Gregor Maurach (Pretoria: University of South Africa, 1974).
[91] William of Conches, *Philosophia mundi*, ch. 15.

range of plant species in the known world, while 16 of the 39 books of Pliny's *Natural History*

addressed plants (for many of these books, Theophrastus is named as a key source). These texts

catalogued medicinal qualities of specific plants, documented plants known to grow in regions

with varying climates, and briefly theorized how soil conditions and weather might affect plants'

growth and fecundity. This branch of plant knowledge infused medieval herbals circulating in

the Middle Ages, which played a similar documentary and referential role.[92] Other procedural

texts about cultivation, such as those by the first-century writer Columella (*De re rustica, De

arboribus*) made up another branch of thought that included practical manuals on agriculture and

forest management, such as Walter of Henley's *Husbandry*, a practical guide to agriculture

written in England, which will be considered in more detail below.[93]

While medieval philosophers stopped short of exploring the subject of plant life in their

treatises, several works of natural philosophy reflect confidence in the ability of "Man" to

control matter and, by extension, "nature." In the intellectual circles of Paris and Chartres, the

desire to understand matter and its ability to transform was a common theme in theological and

philosophical texts.[94] While this confidence might seem integral to the Christian worldview

---

[92] Minta Collins, *Medieval Herbals: The Illustrative Traditions* (Buffalo: University of Toronto Press, 2000); Peter Dendle and Alain Touwaide, eds., *Health and Healing from the Medieval Garden* (Woodbridge: Boydell & Brewer, 2008); Givens, *Observation*, 14, 23, 145-146. The study and development of medieval herbals is a fascinating complement to foliate sculpture. Herbals copied and illustrated in monastery scriptoria across Europe had some features in common with modern botany and the development of plant symbolism, but it is important to note that this genre of manuscript production had its own conventions and complexities quite separate from the production of architectural sculpture.

[93] Walter de Henley, *Walter of Henley's Husbandry: Together with an Anonymous Husbandry, Seneschaucie, and Robert Grosseteste's Rules* (New York: Longmans, Green, and co, 1890). Henley's text is organized by a set of common problems encountered in running an agricultural enterprise.

[94] For an overview of how medieval writers thought about change and the transformation of matter, see Caroline Walker Bynum, "Matter and Miracles," in *Christian Materiality: An Essay on Religion in Late Medieval Europe* (New York: Zone Books, 2011), 217–65, esp. 256-259. Stephen Murray addresses this phenomenon in an architectural context; see Stephen Murray, *Notre-Dame, Cathedral of Amiens: The Power of Change in Gothic* (New York, NY, USA: Cambridge University Press, 1996). Historians have been applying an ecological lens to medieval efforts to control matter and landscape since the publication of Lynn White, Jr.'s article on the destructive nature of medieval Christian thought. This essay has been received critically among scholars of religion and the history of science, but less often among art historians. For a critical

outlined in the pages of Genesis, this is a notable development and even a contrast from the writings of the early Christian fathers. In fact, many early Christian writers had a pessimistic view of nature expressed through apocalyptic anxiety that Earth's resources were being exhausted by overpopulation. Historian David Herlihy noted that third-century Christian writers such as Tertullian lamented over an exhausted landscape ("Everything has been visited, everything known, everything exploited..."). Similarly, St. Cyprian took an apocalyptic view of the use of natural resources and the passage of the seasons:

> You ought to be aware that the age is now senile...There is a diminution in the winter rains that give nourishment to the seeds in the earth, and in the summer heats that ripen the harvests. The springs have less freshness and the autumns less fecundity. The mountains, disemboweled and worn out, yield a lower output of marble; the mines, exhausted, furnish a smaller stock of the precious metals: the veins are impoverished, and they shrink daily...When a thing is growing old, do you suppose that it can still retain, unimpaired, the exuberance of its fresh and lusty youth.[95]

The writings of the twelfth and early thirteenth centuries, on the other hand, largely reflected confidence in the natural world and the possibility of discovery, rather than apocalyptic anxiety or doubt. Nature, it seemed, could be understood, decoded, and mastered, as evidenced by the medieval herbals, bestiaries, and treatises on natural philosophy. For nearly a century, these two strands of ideas, the encyclopedic and the philosophical, would remain distinct from one another. Albertus Magnus' *De vegetabilibus,* composed at the end of the thirteenth century, would bring these two traditions together, the first written work in the medieval West to apply the art of sharp

---

response to White's essay, see Bron Taylor, Gretel Van Wieren, and Bernard Daley Zaleha, "Lynn White Jr. and the greening-of-religion hypothesis," *Conservation Biology* 30, no. 5 (2016): 1000–1009; Anna Peterson, *Religion and Ecological Crisis: The "Lynn White Thesis" at Fifty* (Routledge, 2016); Timothy J. Burbery, "Ecocriticism and Christian Literary Scholarship," *Christianity and Literature* 61, no. 2 (Winter 2012): 189–214.

[95] David J. Herlihy, "Attitudes Toward the Environment in Medieval Society," in *Historical Ecology: Essays on Environment and Social Change*, ed. Lester J. Bilsky, National University Publications (Port Washington, N.Y: Kennikat Press, 1980), 104.

observation to individual species and consider universal laws governing plant growth and
regeneration.

2.3 From *De plantis* to Albertus Magnus: Theories on Plant Growth and Regeneration

Despite interest in plant cultivation and agricultural practice, most natural philosophers
chose not to probe questions related to plant regeneration and growth. This lacuna in medieval
texts left plenty of space for mystery in the understanding of the plant world and, notably, stands
in contrast to the proliferation of sculpted plants in sacred space produced by medieval artisans.
When theories of plant regeneration and leaf formation began to circulate more widely at the end
of the thirteenth century and into the fourteenth, the theories resembled contemporary
explanations of the human body. Albertus Magnus' work *De vegetabilibus* began as an expanded
version of *De plantis;* like his contemporaries, Albertus attributed *De plantis,* which had been
translated into Latin from the Greek by Alfred of Sareshel, to Aristotle. Though we now believe
this text to be the work of Nicholas Damascus from the first century B.C., Albertus and other
medieval writers organized their own work around the assumption they were working with
Aristotelian ideas. For the sake of this brief study, we will leave aside the issue of authorship of
De plantis to contemplate the contribution of Albertus's text.[96]

De vegetabilibus is organized into seven books. Much of the text is devoted to glossing,
reorganizing, and expanding *De plantis* while providing a reference for the qualities of individual

---

[96] On the authorship and translation of *De plantis,* see Ernst H. F. Meyer, *De plantis labradoricis libri tres*
(Lipsiae: sumtibus L. Vossii, 1830), and S. D. Wingate, "The Mediaeval Latin Versions of the Aristotelian
Scientific Corpus: With Special Reference to the Biological Works" (London, The Courier Press, 1931). E.S.
Forster translated Meyer's *De plantis* into English in 1913, published in *The Works of Aristotle Translated into
English:* Vol. 6, *Opuscula,* ed. W. D. Ross (Oxford, 1913).

plants, not unlike medieval herbals or Pliny's *Natural History*. In terms of the reception of Albertus Magnus' works, the study of botany in the nineteenth century enlivened interest in *De vegetabilibus*, but the foundational study written by botanist Ernst H. F. Meyer focused on the sixth book, which addresses individual plants. Parallel trends in the study of foliate sculpture arose during in the nineteenth and early twentieth centuries, with art historians teaming up with botanists to categorize and classify the types of plants they saw sculpted in the jambs, tympana, and capitals of medieval cathedrals, as A.C. Seward did famously at Southwell and E.J. Woillez and Virgile Brandicourt did at Amiens.[97] Albertus was specific and florid in his descriptions of individual plants, a thrilling find for historians of science. From his text, it seems that he not only carefully considered *De plantis* and other sources but made his own observations from nature. In the scientific gaze of mid-nineteenth century Germany, Meyer's incisive essays formed a cornerstone of scholarship on Albertus Magnus as one of the proto-scientists of the medieval period.

And yet, the other six books of *De vegetabilibus,* which do not resemble botanical catalogues in any sense, contain numerous additional insights. Karen Reeds pointed to the importance of books 1-5, which contain Albertus's explanation and expansion of *De plantis* in books 1 and 4 as well as his "digressions" in books 2, 3, and 5, ideas that had not received as much critical attention.[98] *De vegetabilibus* addresses, importantly, the issue that the neo-Platonists had been circling around for decades without addressing its specific application to

<hr>

[97] A. C. Seward, *The Foliage, Flowers and Fruit of Southwell Chapter House. An Original Article from the Proceedings of the Cambridge Antiquarian Society, 1933-1934.*, First Edition (Cambridge Antiquarian Society, 1935); Virgile Brandicourt, "La faune et la flore de la cathédrale d'Amiens: lecture faite à la séance publique du 21 décembre 1905, de la Société des Antiquaires de Picardie," *Bulletin de la Société des antiquaires de Picardie*, 1905; E. J. Woillez, "Iconographie des plantes aroïdes figurées au moyen âge de Picardie," *Mémoires de La Société des Antiquaires de Picardie* 9 (1848).

[98] Karen Reeds, "Albert on the Natural Philosophy of Plant Life," in *Albertus Magnus and the Sciences: Commemorative Essays 1980*, ed. James A. Weisheipl, Studies and Texts - Pontifical Institute of Mediaeval Studies 49 (Toronto: Pontifical Institute of Mediaeval Studies, 1980), 341–54.

plant life: the question of the "soul" of plants, their life source, means of reproduction, as well as their diversity. In Albertus's view, the soul was a crucial issue because without seeking to understand it, "we could not understand the bodies of plants."[99] Reeds highlights, furthermore, that Albertus's intellectual journey into the diversity of plants was not a project of taxonomy; rather, his observations of individual plants served a larger goal of discussing the philosophy of plant life. It is this key intellectual difference that put Albertus's works somewhat at odds with nineteenth-century botanists and intellectuals who did not comment on the value of his overarching philosophy of plant life, lest it lie outside a scientific, taxonomic worldview.

The value, however, of considering Albertus's ideas is not only to seek the existence of certain threads of scientific thought in the medieval period, but to show his intense engagement with plant propagation, variety, reproduction, and his contemplation of their existence. Unique among Albertus's many contributions was his theory of leaf formation. Whereas *De plantis* is sparing in its details surrounding explanations of why leaves were shaped differently—the discussion is limited to narrow and broad-shaped leaves, an explanation that is repeated in Pliny's *Natural History*[100]—Albert expands this idea, describing the plant as a body with veins and humors. Pointed leaves, he wrote, are created from a lack of water and abundance of heat, drawing moisture away from the veins and creating serrated, pointed shapes. Rounded leaves, however, had an abundance of water, a result of a "good" mixture. Albert uses these explanations to explain the varying leaf shapes of several species of plants and trees, applying theories of balance of wet and dry, hot and cold.[101]

---

[99] Qtd. in Reeds, "Albert," 347.

[100] Pliny the Elder, *Pliny's Natural History. In Thirty-Seven Books,* vol. 3, Book 16, Ch. 38, (London: Printed for the Club by G. Barclay, 1847), 377.

[101] Albertus Magnus, *De vegetalibilibus,* Book II, tr. 2, c. 2, para 100-109 (pp. 142-146)

De vegetabilibus introduced not only the combination of the form of a plant catalogue with a broader philosophy of the life of plants, but also an awareness of the plant as a living body with a generative force distinct from but not unlike that of animals and humans. This pointed awareness and curiosity about plant life is notable, as plants had not figured into grand narratives about natural philosophy a century earlier. The Aristotelian text provided Albertus with scaffolding to communicate these ideas, but his synthesis was novel. Though Albertus was aware of individual specimens and applied methods of close observation to support his ideas, he was also a generalist concerned with larger questions of life, *animus*, and regeneration.

Similarly, artists approached the natural world in a variety of ways, using plant representation flexibly to accommodate various interpretive and artistic goals. In architectural sculpture, masons used observation to reproduce nature more frequently when the artwork or iconographic program called for a specific plant with identifiable characteristics. When representing a specific text, for example, the sculptors who created the sculptural program at Notre-Dame of Amiens were careful to reproduce specific species of plants. In the central portal, plant symbolism accompanies the trumeau sculpture of Christ, the Beau Dieu. Below the sculpture, a sturdy vine grows out of the base of the relief, forming a spiral and separating into two branches of the vine with symmetrically placed leaves and small, circular bunches of grapes (fig. 2.1). While the overall composition is symmetrical, the leaves are highly naturalistic, each with serrated edges and carefully incised veins that further define its five-point shape. This vine is complemented by a potted lily and a potted rose flanking the relief of Solomon, both of which grow vertically along the side of trumeau and are only visible when walking around the whole sculptural ensemble (fig. 2.2). The three beautifully rendered plants, taken together, illustrate Song of Songs 2:1 "I am the vine, I am the rose of Sharon, I am the lily of the valley." In representing a specific biblical reference, these sculpted plants play directly into the iconographic ensemble. In other instances, however, such as the heavily restored vine friezes that frame the registers of the central tympanum at Amiens, the diverse foliate friezes that remain at Noyon, the upward-stretching crockets in many French and English capitals, and the swirling vegetal tympana that adorn the western façade and triforium walls of Wells cathedral, the sculpture contains no specific reference to a particular plant. In this way, masons and sculptors worked in both modes of reality and fantasy, natural and super-natural.

2.4 Man over Matter: Plant Cultivation and Climate

It has been argued elsewhere that the economic growth and technological advancements of the twelfth century contributed to a certain confidence in man's ability to control nature and matter, and that this confidence is reflected in a number of written works. Marie-Dominique Chenu wrote that this "new sensibility" surfaced in intellectual circles as well as in daily life, writing that the "encounter between man and nature becomes complete only when man has subdued nature to his service; the order of nature demands man's mastery and, for the Christian, so too does the command given man by the Creator on the first pages of the Bible. The twelfth century followed this command."[102] Rather than view this advancement as a sign of coming apocalypse as many early Christian thinkers did, twelfth-century writers embraced rhetoric, discovery, observation, as well as the codification of technical knowledge.

Our understanding of sculpted foliage has been informed by historical assessments that outline evidence for man's growing confidence in the mastery of nature. Chenu (despite his

---

[102] Chenu, *Nature,* 38

apparent lack of interest in sculpted foliage) highlights Lewis Mumford, the historian of

technology, who also saw foliate sculpture as a sign of growing awareness of the natural world:

> By a slow and natural process, the world of nature broke in upon the medieval dream of hell and paradise and eternity. In the fresh naturalistic sculpture of the thirteenth-century churches one can watch the first uneasy stir of the sleeper, as the light of morning strikes his eyes. At first, the craftsman's interest in nature was a confused one. Side by side with the fine carvings of oak leaves and hawthorne sprays, faithfully copied, tenderly arranged, the sculptor still created strange monsters, gargoyles, chimeras, legendary beasts. But the interest in nature steadily broadened and became more consuming.[103]

While these assessments may hold less weight in recent scholarship, they represent the lineage of

sculpted foliage in teleological arguments that view medieval culture as a foil to the Italian

Renaissance and Enlightenment, an idea that is so entrenched in the study of Western cultural

history that it requires constant prodding and readjustment, even if it is obvious to historians

working in the field today. As studies of medieval religious practice and medieval philosophy

show, the "dream of hell," or state of irrationality, does not simply give way to a renaissance of

science and naturalism; conflicting ideas could, and did, coexist. They had geographic specificity

as well as a material specificity: the interest in sculpting the natural world cannot be explained by

the interest in writing about the natural world on the part of medieval intellectuals, though these

trends do occasionally intersect.

The field of historical climatology has begun to open new avenues of inquiry in the study

of medieval culture, history, and the history of art. In addition to written sources, we can now

look to paleoscientific data made available through dendrochronology, ice cores, and sediment

deposits in lakes.[104] These findings confirm, broadly speaking, two distinct climate periods

---

[103] Lewis Mumford, *Technics and Civilization* (London, 1934), p. 28-29. Qtd in Chenu, *Nature*, p. 39.
[104] For an overview of this research, see Richard C. Hoffman, "Homo et Natura, Homo in Natura: Ecological Perspectives on the European Middle Ages," in *Engaging with Nature: Essays on the Natural World in Medieval and Early Modern Europe*, ed. Barbara Hanawalt and Lisa J. Kiser (Notre Dame, Ind: University of Notre Dame Press, 2008), 11–38. Historical climatologists Glaser, Pfister, and their collaborators have

before what is known as the Early Modern "Little Ice Age": a general phase of cooler climate

trends that coincides with the late antique and early medieval periods, and the "medieval warm

period," which lasted approximately from the tenth through to the thirteenth century. It is

important to note that these climactic patterns had different implications for different

geographical areas. In northern France and England, for example, the "medieval warm period"

corresponded to more dramatic contrast in seasonal changes: warmer summers but also cooler

winters with higher amounts of precipitation.

Just as Pliny and early Christian writers composed their works within the ecological and

cultural milieu of the Mediterranean climate, the practical knowledge that was afforded in terms

of agriculture and the care of plants penned in northern France and England was specific to these

climactic regions—and in particular, these regions as they were experienced in this period, with

heavier precipitation and sunny summer seasons that accommodated and perhaps even facilitated

expanding demand for agricultural production. We can speculate whether this seasonality might

have also contributed to discourses related to the mastery over nature, the desire to engender

confidence to alter landscapes, and shape matter into an infinite number of useful tools, shelters,

and objects. As the didactic tone and participatory inquiry of Walter of Henley's *Husbandry*

suggests, readers were given the opportunity to join the narrator in their contemplation of

common struggles related to agricultural production: "And will you see how the horse costs more

than the ox? I will tell you. [Io le vos dyray.][105]" After a show of empathy and understanding, the

narrator provides a clear solution ("Coment?" he asks, "Io vos dirray"). Such a promise in being

gathered data from continental Europe, Mediterranean Italy, and the northwestern Atlantic in efforts to
corroborate climate trends in the Middle Ages.
[105] Walter de Henley, *Husbandry*, 11.

able to learn how to control the landscape might explain the volume's successful dissemination and multiple translations.

In a similar vein, economic historians have made much of the forest clearances of England and continental Europe during the twelfth and thirteenth centuries.[106] Routine and often sustainable deforestation practices, however, had been commonplace since prehistoric and Roman times. By the twelfth century, forests were already seen as important economic resources, protected by individuals, city communes, or cathedral organizations, as medieval legal texts and charters illustrate.[107] The detailed records kept by cathedral organizations and priories show that land use for agriculture and husbandry was monitored closely; the records from Norwich Cathedral Priory, for example, document actively observant land usage processes from the mid-thirteenth century until the Black Death.[108] Many villages and priories practiced itinerant farming to keep soils productive and fertile, cutting down forests for agricultural fields then allowing them to return to fallow for two or three decades before using them again. Evidence from manorial accounts, inventories, and—increasingly—soil composition analyses suggests that the practice of three-course crop rotation was common in regions of England, northern France, and the Low Countries as early as the thirteenth century, diversified agricultural practices responding

---

[106] Georges Duby, *L'économie rurale et la vie des campagnes dans l'occident médiéval: France, Angleterre, Empire, IX-XV Siècles* (Paris: Aubier, 1962); John Aberth, *An Environmental History of the Middle Ages: The Crucible of Nature* (New York: Routledge, 2012), 93-94. The so-called *"grands défrichements"* thesis originated in the work of Georges Duby in the 1960s and went largely unchallenged for decades. Scholars have since considered the idea that Cistercian foundations led large campaigns of land reclamation and deforestation to be limited, noting that much of this land was already cultivated before these foundations began to spread across Europe and deforestation had been practiced at least since Roman times. More research is needed to understand a more complete picture of how landscape use and alteration might have changed in the twelfth and thirteenth centuries.

[107] Oliver Rackham, "Medieval Countryside," in *Trees and Woodland in the British Landscape* (London. Dent, 1976), 15-19.

[108] E. Stone, "Profit-and-Loss Accountancy at Norwich Cathedral Priory," *Transactions of the Royal Historical Society* 12 (1962): 25–48

to ecology as well as demand.[109] Furthermore, land clearances were driven by agriculture as well

as the need to fuel furnaces, build scaffolding, construct cities, and build ships. Charles Bowlus

estimated that it may have taken 100 acres of woodland just to produce the fuel needed to

produce one square meter of stained glass.[110]

In Picardy, forest clearings peaked in 1210-1220, holding steady until 1240 (and

corresponding with the first phases of construction of Notre-Dame of Amiens).[111] Among the

driving forces for land clearances was the demand for wine; in the twelfth and thirteenth

centuries, vineyards were planted in northern France to respond to ongoing demand. The

enterprise required not only vast expanses of land, but also a wealth of tools and oak barrels for

storage. What is not clear, however, is if this expansion of wine production was also aided by the

warmer summers in northern France during the medieval warm period. Further research is

needed to determine if the placement of new vineyards and their productivity aligns with what is

known about the medieval climate from paleoscientific data, a difficult task in this period due to

the relative scarcity of economic records before 1200 as compared with those from the late

thirteenth and fourteenth centuries. Dioceses were involved in this expansion and benefitted from

its profits; through its land holdings, a cathedral organization could control the production of

---

[109] See, for example, the summary of agricultural land use practice and bibliographies in Bruce M. S. Campbell and Mark Overton, "A New Perspective on Medieval and Early Modern Agriculture: Six Centuries of Norfolk Farming c.1250-c.1850," *Past & Present*, no. 141 (1993): 38–105; Sabine Karg, "Plant Diversity in Late Medieval Cornfields of Northern Switzerland," *Vegetation History and Archaeobotany* 4, no. 1 (1995): 41–50; Corrie C. Bakels, "Crops Produced in the Southern Netherlands and Northern France during the Early Medieval Period: A Comparison," *Vegetation History and Archaeobotany* 14, no. 4 (2005): 394–99.

[110] Charles R. Bowlus, "Ecological Crisis in Fourteenth Century Europe," in *Historical Ecology: Essays on Environment and Social Change*, ed. Lester J. Bilsky, National University Publications (Port Washington, N.Y: Kennikat Press, 1980), 95. The production of stained-glass windows, architectural design, and climactic shifts might relate to regional trends and even weather patterns, though more research is needed. See C. T. Simmons and L. A. Mysak, "Transmissive Properties of Medieval and Renaissance Stained Glass in European Churches," *Architectural Science Review* 53, no. 2 (May 1, 2010): 251–74; "Stained Glass and Climate Change: How Are They Connected?," *Atmosphere-Ocean* 50, no. 2 (June 1, 2012): 219–40.

[111] See chart on p. 311, Robert Fossier, *La terre et les hommes en Picardie jusqu'à la fin du XIIIe siècle*, vol. 1 (Paris, Louvain: B. Nauwelaerts, 1968).

wine by on these lands; Abbot Suger describes several instances in which the Abbey of Saint-

Denis oversaw the financing, cultivation, and wine production of such vineyards, such as this

example in the neighborhood surrounding Saint Denis:

> Besides, near this same place, namely at Saint-Lucien, because our church had an urgent
> need, we had planted and cultivated at great cost a vineyard of about 80 arpents, so they
> say, to which we have assigned for the great profit of the church these same 20 pounds so
> that it will be well used: arrangements well considered, because too often in many places,
> and even at Lagny, because of the lack of wine, the cross, the chalices and the vestments
> are pawned.[112]

Such vineyards could also be established at new foundations financed by the Abbey. Following

the miraculous cure of a woman sick with dropsy, Abbot Suger describes in great detail the

commemorative act of an abbey foundation, including the establishment of a new vineyard:

> We had a vineyard sufficient for a great abbey planted and had acquired by diverse
> means a great number of vines at a good price for them; after this we constructed for this
> same place, without excessive cost, in all propriety, four presses, each with the capacity
> of nearly 80 muids of wine, furnishing this in such abundance that they received amply
> from 250 to 300 muids of wine. On our own nearby domains we had sufficient meadows
> enclosed, and we had suitable gardens prepared to receive the planting of legumes.[113]

It is unclear whether these vineyard projects continued into the fourteenth century, when the

climate became cooler and subject to year-round precipitation, a shift that contributed to a

number of changes and important in the fourteenth century or even in the late decades of the

thirteenth. A sharp decrease in forest clearances in Picardy began in 1280-1300 and continued a

steady decline into the first decades of the fourteenth century.[114] This specific instance and

similar patterns in other locales in northern France and England is often linked to demographic

shifts, but it also happens to correspond to the gradual onset of cooler climactic trends. Indeed,

the fourteenth century would bring years of year-round, wet conditions north of the Alps,

---

[112] Abbot Suger, *De administrione*, trans. L.A. Harkey. *Medieval Sourcebook*, 4-5.
[113] Suger, *De administrione*, trans. Harkey, 17.
[114] Fossier, *La rerre et les hommes*, vol. 1, 311.

endangering crops and contributing to one of the first humanitarian and ecological disasters of

fourteenth-century Europe: the Great Famine of 1317-1320.[115] This event, rather than the

translation and dissemination of Aristotle in the most elite intellectual circles of northern France

and its surrounding regions, appears to be one of the most crucial factors driving the paradigm

shift in how medieval culture viewed nature. More than Aristotle, the Great Famine touched

nearly every segment of the population. The failed crops in the first decades of the fourteenth

century and ensuing famines destabilized political balances of power—and, broadly speaking,

shook the confidence that nature could be decoded and controlled through study and rhetoric.

2.5 Interpretation: Natural Philosophy, Climate, and Sculpted Foliage

While foliate sculpture is rarely commented upon in medieval written sources, it sits

dynamically within the rhetoric of creation found in prominent works of natural philosophy that

take after Chalcidius' commentary on the *Timeaus,* which states that all "things that exist are the

work of God, or the work of nature, or the work of a human artisan imitating nature."[116]

Honorius of Autun, William of Conches, Alan of Lille, Albertus Magnus, and others identified

and expanded upon a rhetorical trinity of creation: God as creator, man as artisan, and works of

nature created by a generative force. The textual record is rich with examples. William of

Conches applies this theory to clothing and architecture:

> The work of an artisan is a work that man engages in because of a need, as making
> clothes for protection against cold or a house against bad weather. But in all that he does,
> the artisan imitates nature, for when he makes clothes, he fashions them after the natural
> disposition of the body's members; and when he makes a house he remembers that water
> that collects on flat surfaces makes wood rot, whereas it flows down off slopes and

---

[115] Centering ecological events in the study of economic and social history suggests that human conflict, wars, and famine were often entangled with changes in climate, weather patterns, and meteorological fallout from natural events such as volcanic eruptions. See Bruce M. S. Campbell, "Nature as Historical Protagonist: Environment and Society in Pre-Industrial England," *The Economic History Review* 63, no. 2 (2010): 281–314.

[116] Qtd. in Chenu, 41.

cleanses them, so he makes his house peaked.[117]

Master Gilbert explained that "the divine creative action, the operations of nature, and the things made by men—all three [are] parts of the divine governance."[118] Similarly, Richard of Saint Victor expanded the idea to several disciplines:

> For the working of nature is one thing and the working of human activity is another. We are easily able to grasp…in trees, how they leaf out, blossom, and bear fruit. A work of human activity is artificial, as in engraving, painting, writing, agriculture … because they co-operate mutually with each other, natural works and artificial works are joined to each other. And are united together in themselves by mutual contemplation. For indeed it is certain that a work of activity takes beginning, continues and gains strength from a natural operation, and a natural operation makes progress from activity, so that is better.[119]

Richard's examples of artificial activity span art-making and agriculture, the physical manipulation of matter, and words. Nature and human activity, he highlights, are intertwined, locked together in a process of growth and human intervention animated by *animus*.

Foliate sculpture fits within this artificial category, as it is the work of the artisan. In Gothic buildings, it is sometimes sculpted into the architectural frame to imitate specific specimens, reflecting movement, change, and regeneration. More than a mere act of imitation or artificiality, there is reason to question whether some examples of foliate sculpture might also reflect a desire to represent *animus*, a generative force represented by tangles of vines that appear to grow directionally or oak leaves situated in perfect, ordered symmetry. A reflexive

---

[117] William of Conches, *Glossa in Timaeum*, in Joseph Marie Parent, *La doctrine de la création dans l'école de Chartres; Études et textes* (Paris: J. Vrin, 1938), 148. Qtd in Herlihy, 112; Chenu, 41.

[118] Master Gilbert, *Notae super Johanne secundum magistrum Gilbertum* (MS London Lambeth Palace 360, f. 32rb). Qtd. in Marie-Dominique Chenu.

[119] Richard of Saint Victor, "The Mystical Ark," in *Richard of Victor: The Twelve Patriarchs, The Mystical Ark, Book Three of the Trinity*, trans. G. A. Zinn (London, 1979, p.180). See also T. A. Heslop, "Art, Nature and St Hugh's Choir at Lincoln," in *England and the Continent in the Middle Ages: Studies in Memory of Andrew Martindale, Proceedings of the 1996 Harlaxton Symposium*, ed. J. Mitchell (Stamford: Shaun Tyas, 2000), 60–74.

consideration of medieval natural philosophies of plants and dynamics of change enrich the interpretive possibilities of this type of sculpture.

The emergence of the idea of a vegetative soul and the rise in a burgeoning genre of practical manuals on plant cultivation prompted new questioning around the lives of plants. Just as the "great craftsman" created the world and its flora, the artisan created and re-made the world through images. Foliate sculpture arrests plant life in a frozen state through the intervention of the mind and the hand of the artisan. In twelfth- and early thirteenth-century sculpture, plants are often depicted with perfect symmetry and precise geometry. Instead of considering this an act against naturalism, these eternal plants represent perfect geometry which, according to Platonic philosophy, is at the heart of the divine. These sculptures are also often quite specific in their references to individual species, despite their symmetry and mathematical execution. This is not always indicative of an ignorance of observation of botanical details, but a desire to perfect those very details. Similarly, plants could be pictured as imperfect, stunted, flawed, or subject to divine will just like their human counterparts, as we will see in the next chapter.

3

Strange Weather:

Environmental Miracles at Notre-Dame of Amiens

Looking to a local manifestation of miraculous foliage and lore, this chapter considers the Invention of the relics of Saint Firmin and the celebration of this feast day in concert with its portrayal on the thirteenth-century tympanum of the north portal, where low-relief depictions of the Labors of the Months, set at eye-level on either side of the portal, frame the sculptural ensemble. In the context of the west façade at Amiens, the miracle of Saint Firmin connects to an interlaced program of sculpted foliate imagery depicting the miraculous, the blessed, and the damned. These examples raise questions about miraculous foliage at Amiens as well as the prevalence of environmental miracle stories in northern Europe, in which saints intervened in periods of drought or famine. At Amiens, the depiction of Saint Firmin's Invention presents an inverted natural order, where unusual weather phenomena and miraculous growth subvert the expectations of seasonal climactic shifts in northern France. Furthermore, this program of sculpted foliage and meteorological phenomena is intertwined pictorially and iconographically with representations of biblical texts, particularly those that relate to proto-apocalyptic events, the Last Judgment, and scenes from the life of the Virgin Mary. At Amiens, there is ample evidence that foliate sculpture is not merely decorative, but gestures towards a larger theological argument positioning the Church as the purveyor of grace and the definitive authority on the miraculous. The intensified foliate sculpture and, in particular, the depiction of ecological and

70

meteorological events will be explored in the context of miraculous happenings sanctioned and

promoted by the diocese of Amiens, especially miraculous weather.

Accounts of environmental miracles, unusual events in which saints' relics were reported

to intervene in meteorological phenomena, are plentiful in the Middle Ages. Abbot Suger

describes two such miracles during the rebuilding of Saint-Denis: a rainy day threatened to delay

the quarrying of stone, but the masons managed to quarry it anyway; a violent storm severely

damages the buildings surrounding the abbey but, thanks to the blessing of Bishop Geoffroy of

Chartres, leaves the skeletal stone structure of the church's unfinished vaults untouched.[120] At

Amiens, at least three environmental miracles are documented. In 1060, for example, the relics

of Saint Honoré were carried around the city of Amiens during a devastating drought, a

procession which resulted in a sudden downpour of rain which prompted the plants to grow.[121] In

the twelfth century, the relics of Saint Firmin were processed around the city during another

drought, which had lowered the river Somme and waters around Amiens, drying out the fields

and contributing to crop shortages. During this procession, a woman recovered miraculously

from her ailments in the presence of the relics and was able to join in the parade.[122] A third

---

[120] "…when the work on the new addition with its capitals and upper arches was being carried forward to the peak of its height, but the main arches---standing by themselves---were not yet held together, as it were, by the bulk of the vaults, there suddenly arose a terrible and almost unbearable storm with an obfuscation of clouds, an inundation of rain, and a most violent rush of wind. So mighty did this [storm] become that it blew down, not only well-built houses but even stone towers and wooden bulwarks. At this time, on a certain day (the anniversary of the glorious King Dagobert), when the venerable Bishop of Chartres, Geoffroy, was solemnly celebrating at the main altar a conventual Mass for the former's soul, such a force of contrary gales hurled itself against the aforesaid arches, not supported by any scaffolding nor resting on any props, that they threatened baneful ruin at any moment, miserably trembling and, as it were, swaying hither and thither. The Bishop, alarmed by the strong vibration of these [arches] and the roofing, frequently extended his blessing hand in the direction of that part and urgently held out toward it, while making the sign of the cross, the arm of the aged St. Simeon; so that he escaped disaster, manifestly not through his own strength of mind but by the grace of God and the merit of the Saints." Suger, *De consecratione*, ed. Panofsky, 108-109.

[121] *Acta sanctorum*, XVI (May III), 611; Josse, La Légende de Sainte Honoré, 16. Qtd. In Gould, "Illumination and Sculpture in Thirteenth Century Amiens," 161.

[122] Jules Corblet, *Hagiographie du diocèse d'Amiens* (Paris: J.-B. Dumoulin, 1869), vol. 2, 166-167.

miracle from the end of the fifteenth century recounts a similar drought procession: the relics were paraded to the church of Saint-Acheul, the original burial place of Saint Firmin according to local tradition, and upon the arrival of the procession at the church, a heavy rain commenced immediately.[123] This practice of processing saints' relics in times of ecological duress was not unique to Amiens or northern France, but the importance of procession in Amiens specifically is reflected in the illustration of relic processions in both the north portal on the west façade and the south transept portal.[124] These examples illustrate that the diocesan authorities framed unusual weather events as requiring saintly intervention. At Amiens, this is especially clear in the lore and pictorial traditions of the local patron saint, Saint Firmin the Martyr.

3.1 Saint Firmin: Sources, Images, and Questions of Authenticity

The development of the cult of Saint Firmin is crucial to understanding this program of sculpture because it provides context around the depiction of the patron saint of Amiens in concert with other the major themes of the façade. Before the start of the construction began in 1220, the twelfth century was witness to several developments in the cult of Saint Firmin, both in the diocese of Amiens and beyond its jurisdiction. These activities informed the composition of the iconographic program of the west façade of Amiens conceived in the 1220s, particularly the north portal, which where the depiction of Saint Firmin, the Invention of his relics, and the environmental miracle during the translation of his relics from his tomb to the cathedral.

---

[123] Corblet, 167.

[124] Cecilia Gaposchkin, "Portals, Processions, Pilgrimage, and Piety: Saints Firmin and Honoré at Amiens," in *Art and Architecture of Late Medieval Pilgrimage in Northern Europe and England*, ed. Sarah Blick (Boston: Brill, 2004), 217–42.

72

Saint Firmin the Martyr, the patron saint of the cathedral and the first bishop of the diocese, was celebrated for founding the first community of Christians in Amiens.[125] Born in Pamplona to a noble family in the third century, it was believed that Firmin was converted to Christianity by Honestus, a disciple of Saint Saturninus of Toulouse. In some accounts, Firmin's father was said to have been baptized by Saint Saturninus himself, presenting a direct spiritual lineage to the apostle Peter, who had baptized Saturninus. This detail was illustrated in one of the reliefs on the lateral screen of the choir, signaling its acceptance into the local lore around Saint Firmin by the end of the Middle Ages. In this tradition, Firmin's connection to early Christian leaders' lineage solidifies his legitimacy as the local apostle of the Christian faith, an argument reinforced both in the fifteenth-century lateral screen as well as the thirteenth-century portal sculpture on the west façade. The quatrefoil reliefs on the south side of the screen underneath the sculpted scenes of the search, Invention, and translation of Firmin's body illustrate several scenes from his life, including his baptism, his education by Saint Honestus, the baptism of his father by Saint Saturninus, his preaching, his anointment by Saint Honestus, his conversion of the Roman senators Arcade and Romule, his activity in Angers and Beauvais, as well as several scenes of saint, in golden-painted bishop attire, healing the sick. The dialogue between the thirteenth century sculpture of the north portal and the fifteenth century sculpture on the south side of the choir screen is worth considering in more detail. While both illustrate the discovery of Firmin's body and the translation of the relics, the fifteenth-century work emphasizes Saint

---

[125] The Acts of Saint Firmin can be traced to two medieval texts, which are thought to have their origins in the early Middle Ages. The first text exists in two versions, the first published by Bosquet in 1636 in *Ecclesaie gallicunae historiae*, edited again by the Petits Bollandistes in 1658, and by P. Stilting in 1760. The second, in a collection detailing the acts of Saturnin, Honestus, and Firmin, was discovered in Florence and edited by Macenda in 1798. The discovery of the body of Saint Firmin is also related by an anonymous author from the 8th century, appearing in a sermon that was prepared to be presented on the day of the anniversary of this ceremony, presumably the feast of the invention of the relics. This sermon was printed in the appendix of Guilbert de Nogent and reproduced, in large part, in the *Annales ecclesiastici francorum*, P. Le Cointe, (vol. 4, p. 182-187).

Firmin's early life and martyrdom. Whereas the quatrefoil reliefs in the north portal outline the months of the year and signs of the zodiac, a foil for the miracle of the unseasonal warmth and blossoming trees associated with Saint Firmin's invention, the fifteenth century work features thirteen medallions illustrating the early life and works of Saint Firmin before his journey to Amiens. This narrative makes a visual argument for the legitimacy of Saint Firmin as a saint in what may have been seen as necessary and more compelling content in the years after the Hundred Years' War.

According to Firmin's *Vita*, Firmin became a priest and at the age of 31 and first went to Beauvais, where he was persecuted by the governor.[126] He then came to Amiens, where he successfully converted many of the locals to the Christian faith and was elevated to bishop of the region. His religious activity drew ire, however, and ultimately Firmin was tried and convicted by Roman political leaders, who imprisoned the bishop and executed him by beheading. According to the text, Faustinianus, a Roman senator recently baptized, recovered Firmin's body and buried it outside of the city. Firmin the Martyr was then succeeded by a bishop of the same name, Firmin the Confessor. The new Firmin constructed a sanctuary at the location of his tomb thought to be located in the rural region of Abladana, southeast of Amiens; this church was first dedicated to the Virgin and two other local martyrs, Saints Ache and Acheul.[127]

In the sixth or seventh century, Bishop Salvius sought to recover the body of Saint Firmin. Unaware of the location of Firmin's burial, Salvius prayed for three days, after which he

---

[126] For Saint Firmin's *vita*, see *Acta sanctorum*, ed. Jean Bolland, Jean Carnandet, et al. (1863; reprinted Paris, 1965), September 7, cols. 24-36. For Firmin's translation, see *Acta sanctorum*, ed. Jean Bolland, Jean Carnandet, *et al.* (1863; repr., Paris, 1965), January 25, cols. 34–5. The *Actes de Saint Firmin* were first published by Bosquet in 1636 in *Ecclesaie gallicunae historiae*, vol. 4, then in a collection edited by les Petits Bollandistes in 1658 (*de sancta Honesto*, XIV febr.) and by P. Stilting in 1760 (xxv Sept, with annotations). See also Corblet, *Hagiographie*, vol. 2, 31-188; Charles Salmon, *Histoire de saint Firmin, martyr, premier évêque d'Amiens, patron de la Navarre et des diocèses d'Amiens et de Pampelune* (Rousseau-Leroy, 1861).
[127] Hubscher, *Histoire d'Amiens*, 49. Saints Ache and Acheul are represented in the portal sculpture of the north portal, framing the trumeau of Saint Firmin.

followed a miraculous light which led him to the exact location of the tomb. Upon opening of the tomb, the body of Saint Firmin released a strong fragrance of diverse flowers, "candida Lilia et versants Rosas, aliosque virentes herbarium flores pulchre." This odor was strong enough that it attracted residents from the neighboring towns of Beauvais, Cambrai, Noyon, and Thérouanne to see the miracle unfold. With his audience increased in size, Salvius presented the body of Saint Firmin, lifting it out of the tomb. As he did this, a "boiling heat" came "into the world," turning the cold winter's day into a mild, temperate spring. As the bishop paraded Firmin's body back to the city of Amiens, his audience following him, the heat prompted trees to burst into bloom and bear fruit. The celebrating laypeople and clergy made wreathes from the leafy branches as they processed back into the city.

While accounts of Saint Firmin date from the early Middle Ages, Saint Firmin's martyrdom was marked on liturgical calendars consistently throughout the diocese. The twelfth century was an especially active period for Saint Firmin's cult: not only were two precious reliquaries commissioned over the course of the century—the second of which was not completed until 1204—textual references to Saint Firmin appear more frequently in manuscripts dating from this period. In a ninth-century sacramentary from Amiens currently in the Bibliothèque nationale de France, the twelfth-century additions in the margins of the original text include masses to be sung on the feast day of Saint Firmin.[128] This addition speaks to the need to update the older text to accommodate liturgical ceremonies concerning the saint (fig. 3.1). A twelfth-century psalter from the abbey of Saint-Fuscien and as well as two twelfth-century manuscripts from Corbie mark Firmin's feast days in the liturgical calendar, while numerous thirteenth century manuscripts from Corbie, Saint-Fuscien, Saint-Acheul, Saint-Quentin, and

---

[128] See notes on this text in V. Leroquais, *Les sacramentaires et les missels manuscrits des bibliothèques publiques de France* (Paris, 1924).

Saint-Martin-aux-Jumeaux mark the saint's day — sometimes only the feast day in September, but often both the feast day and the Invention celebrated in January.[129] The late-thirteenth-century ordinal from the cathedral chapter documents five feasts celebrating Firmin: in addition to his martyrdom and invention, the ordinary also marks the octave of the martyrdom, Firmin's arrival in the city of Amiens, and the placement of the relics in the new reliquary.[130] Among the liturgical practices related to Saint Firmin in the ordinal, the re-creation of the miraculous winter heat is especially notable: in this ceremony in the middle of January, canons would change into their summer robes and don foliate crowns, bestowed upon them by the *homme vert*, to celebrate the invention.[131]

Within the diocese and beyond, the cult of Saint Firmin spread through the dispersion of relics, liturgical ceremonies, and, in some cases, religious foundations. In 864, Bishop Hilmerade offered a fragment of Saint Firmin's tunic to Oudulphe, the treasurer of Saint-Riquier.[132] Later in the ninth century, Bishop Otger offered relics of Saint Firmin to the collegiate church of Saint Quentin, presumably because he had once been a canon there himself. In the eleventh century, a piece of Saint Firmin's arm was given to Alix de Crespy, a woman whose cousin Foulques was then bishop of Amiens, placing her relic in the crypt of a church in Provins. Another eleventh-

---

[129] In the Bibliothèque municipale d'Amiens, the following manuscripts originating from the diocese note the celebrations of Saint Firmin in the liturgical calendar: Ms 20, Ms 111, Ms 112, Ms 124, Ms 156, Ms 149. At the Bibliothèque de l'Arsenal, in Paris, a psalter from Saint-Quentin also includes the feast day of Saint Firmin. "Psaultier à l'usage de Saint-Quentin," Ms 120, Bibliothèque de l'Arsenal, Paris. See also Cambrai, Bibliothèque municipal Ms. 38, which contains a chant on the Invention on folio 412r, as well as an antiphon on folio 343r. Both mention the flowering of the miracle: "Cumque diu talinus floreret auspiciis ad Christum populous convertit pietate suprema."

[130] *Ordinaire de l'église Notre-Dame, cathédrale d'Amiens*, ed. Georges Durand (Amiens: Société des antiquitaires de Picardie, 1934).

[131] See Murray, *Notre-Dame of Amiens* (1996), 115.

[132] Corblet, 164.

century donation included Saint Firmin's rib, which was given to the collegiate church of Saint-Martin de Picquigny.[133]

In the thirteenth century, it was not only the holy matter of Saint Firmin that was shared among Christian communities, but also liturgical practices. In 1229, Thibaud d'Amiens, who had become archbishop of Rouen, established a celebration of Saint Firmin in his cathedral. This liturgical exchange was mirrored in Amiens, where the clergy celebrated the feast of Saint Romain in turn.[134] Events celebrating Firmin continued to be established in various locations into the fifteenth century and beyond.[135] A number of parish churches, collegiate churches, chapels, and altars were dedicated to Saint Firmin throughout France and Spain. Within Amiens, such foundations included the parish church Saint-Firmin-en-Castillon, which played an important role in liturgical processions within the city: it was from this church that the cleric dressed as the *homme vert* would be chosen for the feast of the Invention of the relics. The twelfth-century church of Saint-Firmin-à-la-Pierre, also called Saint-Firmin-à-la-Porte and Saint-Firmin-au-Val, was reconstructed in the sixteenth century and destroyed in the French Revolution, while in nearby Abbeville on the current site of Saint Vulfran, a chapel was dedicated to Saint Nicholas and Saint Firmin. In and around Paris, the first chapel on the north side of the ambulatory at the Abbey of Saint Denis was also dedicated to Saint Firmin, a point to which we will return. The Collégiale Saint-Firmin de Montreuil was established in the twelfth century by Thibaud d'Heilly, bishop of Amiens, one of many religious foundations that was suppressed during the French Revolution, as well as the Séminaire Saint-Firmin de Paris on rue Saint-Victor in Saint Denis, converted to a prison in the Revolution. In 1216, Renault d'Amiens, signeur of Vignacourt,

---

[133] Corblet, 164.
[134] Corblet, 171.
[135] For example, the fifteenth-century archdeacon of Notre-Dame of Paris Jean de Courcelles, a native of Amiens, established an annual procession on Saint Firmin's feast day. See Corblet, 171.

founded a chapter of Saint Firmin the Martyr which would remain active until the eighteenth

century.[136] Finally, in Pamplona, Spain, a chapel was dedicated to Saint Firmin as was a spring

near the church of Saint Saturnin, associated with the first baptisms in the city. These activities

show the many ways in which the cult of Saint Firmin was promoted through campaigns of relic

gifts, liturgical practices, and dedications.

In addition to foundations bearing the saint's name, Saint Firmin's image appeared in

sculpture and stained glass, of which few examples survive. The earliest known depiction of

Saint Firmin in sculptural form can be found in the north portal on the west façade of Amiens, in

the trumeau figure of Saint Firmin and the depiction of the Invention in the tympanum above.[137]

Saint Firmin also appears in the stained glass of the cathedral in the axial bay of the glazed

triforium, where he and John the Baptist flank a scene of the Annunciation (fig. 3.2). This stained

glass was executed around 1260 and reflects the devotional priorities of the cathedral, depicting

the saints whose remains accounted for its most important relics.[138] In the stained glass, Saint

Firmin is shown dressed in his bishop's robe and miter, holding his hand in a gesture of blessing,

his robes a vivid green. In its pose and gesture, the image has much in common with the trumeau

sculpture in the north portal on the western façade, which depicts Firmin in a similar stance,

echoing the posture of the *Beau Dieu* trumeau in the central portal.

It is the environmental miracles related to the invention of Saint Firmin's relics, however,

that are illustrated in the most vivid detail, both in the north tympanum as well in a series of eight

---

[136] Corblet, 170.

[137] Several later statues of the saint exist in and around Amiens, including in the sculpture of the parish
churches Saint-Germain, Saint-Leu, and Saint Firmin, Saint Vulfran in Abbeville, as well as several other
churches in the diocese, including Frémont, Herbécourt, Saint Riquier, and the Chapelle du Saint-Esprit in
Rue. Other statues outside of the diocese include examples at Sommesnil and Pamplona Cathedral. There are,
of course, many other statues that have disappeared, including examples from Saint-Firmin-à-la-Pierre.

[138] The north side of the clôture features sculptural scenes depicting the life and martyrdom of Saint John the
Baptist.

monumental sculpted scenes on the south lateral screen of the *clôture* (lateral screen)

commissioned in the late fifteenth century, separating the liturgical center of the choir from the

flanking aisles. Among the destroyed images of Saint Firmin for which we have documentation,

the fifteenth-century stained glass and sculpture at Saint-Firmin-en-Castillon is notable: a series

of stained glass images illustrated the life of Saint Firmin, while a fifteenth-century sculpture of

Saint Firmin was situated in a niche decorated with flowers and fruits, recalling the

environmental miracle of the Invention.

The other known image of Saint Firmin from the thirteenth century, a full-page

manuscript page from the Hours of Yolande of Soissons completed around 1280-90, depicts not

the martyrdom or scenes from his *vita*, but the Invention of the saint's relics. The book of hours

features in its opening folios a full-page illustration of Saint Salvius discovering the sarcophogus,

from which plants are sprouting along the bottom of the page and above the head of the saint.

Three orange rays shine down from the heavens towards the body of Firmin, indicating the

position of the tomb's burial (fig. 3.3).[139] This image, shown in concert with an image of Saint

Francis on the following folio, is an intriguing inclusion in this book of hours; the reception of

Saint Francis and Franciscan thought in Amiens is evident earlier in the thirteenth century, as

Saint Francis appears in the central tympanum of the west façade as the first of the elect to enter

heaven, one of the earliest representations of the saint.[140] Other books of hours from Amiens

date from the 14th or 15th century and do not include images of the local saint—or Saint

---

[139] Barnes Carl F., "Cross-Media Design Motifs in XIIIth-Century France: Architectural Motifs in the Psalter and Hours of Yolande de Soissons and in the Cathedral of Notre-Dame at Amiens," *Gesta* 17, no. 2 (January 1978): 37–40; Karen Gould, "Illumination and Sculpture in Thirteenth-Century Amiens: The Invention of the Body of Saint Firmin in the Psalter and Hours of Yolande of Soissons," *The Art Bulletin* 59, no. 2 (1977): 161–66.

[140] In some liturgical calendars from the diocese of Amiens, the feast days of Saint Francis were added at a later date, including the twelfth-century psalter from Saint-Fuscien, Amiens BM Ms 19 and a fifteenth-century book of hours from Amiens, Amiens BM Ms 206. The calendar in the book of hours is written in a Picard dialect and includes the added feasts of "Translacio S. Franciscy" and "Signacula S. Francicy".

Francis—though the feasts of Saint Firmin are included in four calendars in books of hours from

Amiens and one from Corbie.[141]

These activities in service of the local saint, especially those in the decades before and

after the construction of the cathedral, also point to concerns over the authenticity of the saint's

miracles and efficacy of the relics to be housed in the elaborate, thirteenth-century shrine to be

kept at the high altar of the choir. And indeed, there might have been reason to secure the

authenticity of the saint and alleviate doubt; written accounts document a legend claiming that

the body of Saint Firmin did not remain in Amiens and was, in fact, translated to the Abbey of

Saint Denis by King Dagobert himself.[142] This account is largely dismissed in hagiographic

studies of Saint Firmin, if it is mentioned at all, as it casts doubt on the legitimacy of the relics

kept in Amiens. At Saint-Denis, however, the memory of this tradition remains in the existence

of the chapel dedicated to Saint Firmin, positioned on the north aisle of the chevet. Guibert of

Nogent, writing around 1125, also notes the controversy in *De sancti et pignoribus*. Though

Guibert's text was not circulated widely, he was very likely in contact with clergy at Amiens, as

his predecessor at the Abbey of Nogent, Godfrey, had been promoted to Bishop of Amiens before

his arrival.[143] His text provides a rich source to consider issues of authenticity and doubt in late

medieval cults of saints—and is particularly relevant to this study, as it specifically mentions the

relics of Saint Firmin and the new châsse commissioned in the first quarter of the twelfth

century.

---

[141] Book of Hours, fourteenth century, Bibliothèque Louis Aragon, Ms 204, Amiens, France, calendar on fol. 2-6; Book of Hours from Corbie, fifteenth century, Bibliothèque Louis Aragon; Ms 200, Amiens, France; Book of Hours, fifteenth century, Bibliothèque Louis Aragon, Ms 206.

[142] Recuil, fifteenth century, Bibliothèque nationale de France, Bibliothèque de l'Arsenal, Ms. 1030, fol. 71, Paris, France.

[143] Thomas Head, note 50, in Guibert of Nogent, "On Saints and Their Relics," in *Medieval Hagiography: An Anthology*, trans. Thomas Head (New York: Routledge, 2001), 417.

Following a discussion of the existence of multiple heads of John the Baptist (the cathedral of Amiens possesses one), Guibert continues to discuss a similar controversy regarding the relics of Saint Firmin:

> When my predecessor, the bishop of Amiens, transferred what he thought to be the body of St. Firminus the martyr from one casket to another, he failed to discover any document inside, not even a single letter of testimony as to who lay there. I have heard this with my own ears from the bishop of Arras, and even from [his successor as] bishop of Amiens. For which reason the bishop forthwith had an inscription made on a leaden plate, which would lie in the reliquary: "This is Firminus the martyr, bishop of Amiens."[144]

Guibert goes on to recount a related episode at the Abbey of Saint Denis:

> Not long afterward, the incident was repeated in a similar manner at the monastery of Saint-Denis. Relics were taken forth from their resting place in order to be placed in a more ornate shrine, which had been prepared by the abbot. When the skull was unwrapped along with the bones, a slip of parchment was found in the martyr's nostrils, on which it was written that this body was Firminus, the martyr of Amiens. Things are not as those from Amiens claimed them to be in this matter, for written testimonies give voice to a contrary claim and reason, if you please, takes the seat of judgment. Will not the inscription placed on that metal plate by the bishop be judged legally null and void? Does his claim become valid merely by being written down? Surely those from Saint-Denis would object, and they at least have [older] writings on their side.[145]

Guibert is referring to legends that tell of the translation of the body of Saint Firmin to the Abbey of Saint Denis. What ensues, according to Guibert, is a conflict of authenticity.

> So we see that those people who venerate a patron saint about whom they are unsure are always in great danger, even if that patron turns out to be ancient. For if that patron is not a saint, they have committed an enormous sacrilege. What is a greater sacrilege than to venerate as holy something that is not?[146]

Whether or not Guibert's account is true, it is notable that he references the text detailing the translation as well as the commission of the new reliquary, the first of two new containers fabricated in the century before the construction of the cathedral. The second container, for which fundraising began at the end of the twelfth century, was completed just fifteen years

---

[144] Guibert of Nogent, "On Saints and Their Relics," trans. Thomas Head, 417.
[145] Guibert of Nogent, 417-418.
[146] Guibert of Nogent, 418.

before cathedral construction began. This information provides background to consider the

iconographic framing of the miracle of the Invention, pictured in the tympanum of the north

portal, as well as the elaborate program of miraculous foliage sculpted on the west façade. What

was at stake for the diocese of Amiens in situating Saint Firmin pictorially in the seat of spiritual

power?

3.2 Sculpted Foliage in the West Portals

Before considering the depiction of Saint Firmin and the role of miraculous weather at

Notre-Dame of Amiens in more detail, it is important to understand the broad, interwoven foliate

themes employed across the three portals. Depictions of the Tree of Jesse, Aaron's Rod, and

Paradise establish connections with biblical stories, while portraits of individual, naturalistic

plants take on symbolic roles. Still more sculpted foliage interacts with lesser-known biblical

references, often with surprising specificity. The foliage in the portals, then, invites engagement

with the uniquely effusive foliate sculpture on the cathedral's exterior and interior.

Foliate sculpture appears on the upper reaches of the elevation, articulating each tower as

well as the central portions of the façade; gables and towers bristle with crockets.[147] Foliate

forms adorn nearly every architectural surface: decorative flourishes occupy the blind trefoil

arches on the pier buttresses while interlocking rinceaux visually unite the front-facing

quatrefoils underneath the front-facing column figures of the prophets, the forms becoming more

complicated and elaborate from right to left as the construction on the façade progressed. Taken

---

[147] During the restoration efforts directed by Viollet-le-Duc and carried out by the Duthoit brothers, hundreds of crockets were re-made and replaced on the exterior of the building, as well as the robust foliate frieze that crowns the King's Gallery on the west façade, Viollet-le-Duc's design. Michel Barjon, ed., "La Cathédrale d'Amiens façade occidentale: les restaurations des XIX et XX siècles" (Direction régionale des affaires culturelles de Picardie, Conservation régionale des monuments historiques, Groupe de Recherche Art Histoire Architecture et Littérature, Paris, 1993), Bibliothèque Louis Aragon, Amiens.

together, these diverse foliate elements produce an effect of vitality: foliate sculpture eases

architectural transitions and provides a textural richness that responds to changing light

conditions. But in addition to its formal role in the architectural ensemble, the sculpted foliage of

the west façade engages the viewer in questions of authenticity of miraculous events and the role

of the Church in defining and sustaining the natural order.

Scholars of Notre-Dame of Amiens have long noted the coexistence of several modes of

vegetal carving in the building. This diversity has been explained as an indication of the division

of labor at the cathedral *chantier* and as a sign of stylistic transition.[148] In some cases, the range

of foliate forms can be explained by ongoing revisions to the building, but changes in taste

cannot explain each variation.[149] Rather than focusing on inconsistency, however, it is worth

considering why sculptors might have chosen to represent plant life in a multiple ways and the

unique role foliate sculpture plays in the sculptural ensemble and architectural frame. Mary

Carruthers notes that expressions of *varietas* can be found in Constantine's Christian basilicae as

well as Carolingian and Romanesque churches, where *varietas* is expressed through variation,

pattern, materials, and images; we might extend this theory to certain Gothic churches in which

variation is an important component of architectural design and especially architectural

decoration.[150] At Amiens, the variety of sculpted plants invites us to attend to this diversity.

---

[148] See Georges Durand, *Monographie de l'église Notre-Dame, cathédrale d'Amiens*, vol. 1, 113; Denise Jalabert, *La Flore sculptée des monuments du Moyen âge en France, recherches sur les origines de l'art français* (Paris: A. et J. Picard et Cie, 1965); and Meredith Cohen, *The Sainte-Chapelle and the Construction of Sacral Monarchy: Royal Architecture in Thirteenth-Century Paris* (New York: Cambridge University Press, 2015), 68-71.

[149] The placement and delicacy of sculpted foliate forms make them particularly vulnerable to erosion and the elements. Much of the foliate sculpture has been updated, replaced, and occasionally reimagined: the foliate frieze above the King's Gallery is largely the invention of Viollet-le-Duc. The vast majority, however, is original to the medieval structure.

[150] Mary J. Carruthers "'Varietas': A Word of Many Colors." *Poetica* 41, no. 1/2 (2009): 20-22.

While the vine at the foot of the Beau Dieu has a clear textual reference to the Song of Songs, other sculpted plants are not explained as easily. For example, the doorjambs of the central portal picture the Wise and Foolish Virgins, the biblical parable from Matthew 25:1-13, echoing the iconography of the Last Judgment pictured in the tympanum above (fig. 3.4). The "wise" virgins, flanked by miniature columns with foliate capitals, hold their lamps full of oil while the "foolish" hold lamps that are empty. On the bottom register on each side, however, the virgins have been replaced by trees (fig. 3.5). The "wise" tree is flourishing, its branches full of foliage, and full lamps hang from its branches. The "foolish" tree is withered, its empty branches barren of foliage and barely able to support its empty lamps; this tree is so weak that it has even been tied to its own architectural frame, as if it may collapse if were not for the rope encircling its trunk. While trees are not part of the original parable, they might reference the parables of the fig tree in Matthew 21:18-22 and Matthew 24:32-35 or perhaps the parables of the good tree and evil tree in Matthew 7:17-18 or Matthew 12:33. The iconographic connection, in this case, is not entirely clear; whereas fruits and specific leaf shapes are depicted with remarkable specificity elsewhere on the façade, these trees do not have apparent figs or fruits. The wise and foolish trees are unique to the sculptural program at Amiens, and do not appear in representations contemporary or prior to the execution of the portal, either in stone or illustrated in manuscripts.[151]

The inclusion of the trees also suggests that it is not only humans who are subject to judgment and grace, but all of God's creations. A withered tree indicates a lack of grace, whereas a tree in bloom reflects abundance and regeneration, similar to the elect rising from their

---

[151] In architectural sculpture, similar illustrations of the parable of the Wise and Foolish Virgins can be found flanking the central portals of Saint-Denis and Notre-Dame of Paris. These examples do not contain any comparative foliate elements.

sarcophagi and entering heaven in the tympanum above. While the presence of the trees reinforces the tropological messages of the central portal, one might also wonder about the inclusion of these trees at the eye-level of visitors to the cathedral. To layfolk, the message might be that not only their souls, but plant life—and, by extension, agricultural output—could be protected through religious devotion, or destroyed for lack of it. This moralistic perspective suggests that foliate sculpture could, in fact, convey meaning; a building that, in following the prescriptions of Christ and the Bible, is literally blossoming with foliate manifestations of grace on virtually every surface, suggests the divinity of the sacred space enclosed within.

The central portal is not the only example in which flourishing plant life reflects favor and grace. In several narrative scenes contained within the quatrefoils below the column figures, plants are not merely decorative scenery, but vehicles of narrative. Two of these quatrefoils in the south portal depict canonical Biblical stories rich with plant symbolism: Moses and the burning bush and the blossoming of Aaron's rod. These two figures also appear flanking the daïs which crowns the trumeau, seated on either side of the Virgin Mary. In the former, which is the innermost quatrefoil on the south side of the portal, Moses stands with his hands held up, astonished at the image of the burning bush which occupies the right lobe of the quatrefoil (fig. 3.6). The other quatrefoil, placed to the right of the burning bush, presents Aaron, standing on the right of the composition holding his blossoming rod (fig. 3.7). The resonance between the burning bush and the blossoming rod is made here pictorially, with both references to plant life occurring in the right lobe of the quatrefoil; where the bush burns, the rod blossoms in its place.[152]

---

[152] Considering the Annunciation scene above, these Old Testament scenes emphasize the fulfillment of prophecies through Christ. Here, the Annunciation is compared to God's appearance to Moses in the burning bush; Moses will lead his people out of slavery into a new land, just as Jesus will deliver his community from sin to spiritual salvation. Aaron's rod blossoms because he, too, is chosen by God; similarly, Mary is chosen

Other reliefs that depict plant life, however, refer to more obscure events from the lives of

the minor prophets, a program of images created specifically for the cathedral.[153] Instead of

following a pre-established iconographic model or looking to prototypes in other media such as

illustrated Bibles, stained glass, or metalwork, it appears that these scenes were uniquely selected

for representation at Amiens. Among the twenty-four quatrefoils that illustrate scenes relating to

the minor prophets in the column figures on the buttresses, seven feature foliage relating to

proto-apocalyptic events described in the Old Testament.

Below the column figure of Joel, one quatrefoil relief represents the prophet between two

withered trees from which individual leaves have fallen (fig. 3.8). Joel stands on the ground, his

hands clasped. The trees are rendered with specificity and occupy the space of the quatrefoil,

much like the figure of Joel, overlapping the inset border of the relief. The vine, depicted on the

right, appears to sprout from the border itself. This quatrefoil illustrates verses Joel 1:7-12:

> He hath laid my vineyard waste, and hath pulled off the bark of my fig tree: he hath
> stripped it bare, and cast it away; the branches thereof are made white. ... The vineyard is
> confounded, and the fig tree hath languished: the pomegranate tree, and the palm tree,
> and the apple tree, and all the trees of the field are withered: because joy is withdrawn
> from the children of men.[154]

Most astounding in the sculpted image are the falling leaves, carved in shallow relief, which are

shown falling from the two withered branches (fig. 3.9, 3.10). Like the examples below the Beau

Dieu, the modeling of these leaves reflects knowledge of plant specificity: the leaf falling from

by God and her womb will "blossom and bear fruit" in the birth of Jesus. These two scenes, along with
Nebuchadnezzar's dream and Gideon's Fleece, point to Old Testament prophecies pertaining to the birth of
Christ. The remaining quatrefoils on this side of the portal report the events of John the Baptist's birth below
the Visitation, and scenes from the early life of Christ below the Presentation in the Temple—with the
exception of the Fall of Idols, the prophecy pertaining to Jesus' arrival.

[153] A. Katzenellenboken, "The Prophets on the West Façade of Amiens Cathedral," *Gazette Des Beaux Arts*,
Series 6, no. XL (1952): 280–90.

[154] All scripture quotations are taken from the Douay-Rheims translation of the Latin Vulgate. *Bible, Douay-
Rheims Translation* (Project Gutenberg, 2005), <www.gutenberg.org>.

the tree is in the shape of a rounded-lobe fig leaf, and the leaf falling from the vine is unmistakably in the shape of a grape leaf with five-point, serrated edges.

The quatrefoil on the lower register beneath Haggai shows four men standing below a tree. Though this relief has sustained damage, the details of the scene are preserved: one man grabs the tree while a fig drops into the mouth of another (fig. 3.11). This quatrefoil illustrates a text from Nahum 3:12: "All thy strongholds shall be like fig trees with their green figs: if they be shaken, they shall fall into the mouth of the eater." This image is presented as part of a prophecy foretelling the destruction of Ninevah; the fortresses of the city that will fall as easy as fresh figs. This destruction of Ninevah is pictured in a relief on the facing buttress, where a quatrefoil of a falling fortress is paired with a relief of three decimated plants under a cosmic sky of stars (fig. 3.12). The three plants reference the three distinct crops mentioned in Haggai, 10-11:

> Therefore the heavens over you were stayed from giving dew, and the earth was hindered from yielding her fruits. / And I called for a drought upon the land and upon the mountains and upon the corn and upon the wine and upon the oil, and upon all that the ground bringeth forth, and upon men, and upon beasts and upon all the labor of their hands.

Here, it seems that the rebuilding of the cathedral of Amiens is likened to the rebuilding of the temple, part of the insurance in guaranteeing the continuing prosperity and fecundity of the surrounding region. If the temple is not rebuilt, God will bring destruction upon the land and its people. While prophecies involving the fecundity of the land are common in the Old Testament, the choice to represent these particular narratives on the thirteenth-century cathedral of Amiens reflects a relationship between belief, sacred space, prosperity, and the productivity of local agricultural activities in this unique local context. Furthermore, this proto-apocalyptic imagery resonates with the barren and flourishing trees below the Wise and Foolish Virgins in the central portal.

Quatrefoils beneath the column figures of Micah and Jonah show figures sheltered by foliage (fig. 3.13, 3.14). Beneath Micah, two men are seated on the outer lobes of the quatrefoil. On the left, a fig tree grows out of the border of the quatrefoil and bends over to shelter the first man, who is giving a fig and receiving a bunch of grapes, while the other figure is seated underneath a twisting vine. In both plants, the leaves are distinct from one another; the fig tree has naturalistic fig leaves, while the vine has distinct grape leaves, illustrating Micah 4:4: "And every man shall sit under his vine, and there shall be none to make them afraid: For the mouth of the Lord of hosts hath spoken." This verse refers to a prophetic vision of peace, seen here as an agrarian image of arbitration through nonviolent means.

In conversation with this image is the quatrefoil illustrating Jonah seated under the shelter of a "gourd," pictured in the relief as a winding, twisting tree with broad, flat leaves and gourds on the branches, which resemble pomegranates (fig. 3.14). A small lizard takes a bite out of the tree's trunk at its root. This relief illustrates Jonah 4:5-7, in which the prophet sits outside the city of Ninevah, which was to be destroyed before God had changed his mind.

> And the Lord God prepared an ivy, and it came up over the head of Jonah, to be a shadow over his head, and to cover him (for he was fatigued), and Jonah was exceedingly glad of the ivy. But God prepared a worm, when the morning arose on the following day: and it struck the ivy and withered.

Here, the foliage represents shelter given and taken away by God, underscoring the importance of divine protection. As in the examples above, this scene emphasizes the role of the divine in controlling the fecundity of plant life, whether it acts as shelter, nourishment, metaphor, or a combination thereof.

3.3 The Labors of the Months

Similar compositional principles are at play in the quatrefoils that illustrate the labors of the months in the portal of Saint Firmin, which provide a prototype for the expected natural order of the seasons with notable specificity. While the quatrefoils examined thus far have referenced specific Biblical texts, two series of reliefs at Amiens illustrate attributes and concepts: the vices and virtues in the central portal and the labors of the months with corresponding zodiac signs in the north portal, the portal of Saint Firmin. While the relationship between the vices and virtues and the Last Judgment in the central portal has been well-understood and carefully studied, considerably less scholarship has been devoted to understanding of the placement of the Labors of the Months that accompany the tympanum showing the invention of the relics of Saint Firmin at Amiens.[155] At the time of writing this dissertation, I am not aware of any explanations or art historical interpretations of this image cycle specific to the portal sculpture at Amiens.

The labors of the months appear so widely in manuscripts and monumental sculpture throughout continental Europe that it can be difficult to differentiate between the examples in such a large corpus of images, which has its roots in Antiquity and was used in numerous contexts across geographic regions. In addition to late Antique examples in pavement mosaics and monumental sculpture, sculpted iterations of the series appeared in architectural contexts across Italy, France, and Germany; the tradition is not as common in English architecture, though early medieval fonts at Burnham Deepdale and Brookland Church depict them and the signs of

---

[155] As Willibauld Sauerländer, Stephen Murray, Bruno Boerner, and many others have noted, the sculptural program of the west façade at Amiens is a complex collection of highly coordinated, interconnected images. Not only does each portal have its own advanced, reflexive iconography, the portals also interrelate to one another across the façade: for example, the visual rhyme of the daïs that surmounts the trumeau of the Mère Dieu in the south portal and the châsse that crowns the trumeau of Saint Firmin, or the column figures of the apostles and the column figures of bishops and local saints.

the zodiac periodically make an appearance in English portal sculpture.[156] The labors were

frequently carved into the architectural frames of portals, appearing in a variety of compositional

formats: sculpted in a series of roundels above the tympanum, in vertical columns decorating

door jambs, in small registers below column figures, or in the archivolts, as at Chartres and

Notre-Dame of Paris. In manuscripts, they made a logical pairing with liturgical calendars; a

thirteenth-century liturgical manuscript from Amiens, for example, includes illustrations of the

labors of the months in roundels with the corresponding months of the year.[157] The resonance

between these attributes of the months and the demarcation of liturgical time is also at play when

the cycle appears in sacred architecture.

While this image cycle clearly appears in many architectural contexts in France, each

sculptural program is unique in its execution, placement, and interaction with surrounding

sculptural representations of saints. When the labors image cycle frames a tympanum, for

example, as at Autun, Vézelay, and Chartres, they function as a cosmological border outlining

the image of an important religious personage at the tympanum's center, usually Christ. At Autun

and Vézelay, the alternating zodiac signs and Labors surround the tympanum depicting Christ

---

[156] Many studies of the labors of the months image cycle have used the Index of Medieval Art (formerly Index of Christian Art) as a starting point. See Colum Hourihane, ed., *Time in the Medieval World: Occupations of the Months and Signs of the Zodiac in the Index of Christian Art*, Index of Christian Art Resources 3 (Princeton: University Park: Dept. of Art & Archaeology, Princeton University ; In association with Penn State University Press, 2007); J. Carson Webster, *The Labors of the Months in Antique and Mediaeval Art to the End of the Twelfth Century* (Evanston, Ill: Northwestern University, 1938). French social historian Robert Fossier used the reliefs of the labors at Amiens to highlight agricultural practices in Picardy in *La terre et les hommes en Picardie jusqu'à la fin du XIIIe siècle* (Paris, Louvain: B. Nauwelaerts, 1968). Many studies of the labors from the 1980s and 1990s focus on images of work and measurement of time: Jacques Le Goff, *Time, Work & Culture in the Middle Ages* (Chicago: University of Chicago Press, 1980); Perrine Mane, *Calendriers et techniques agricoles : France-Italie, XIIe-XIIIe siècles* (Paris: Le Sycomore, 1983); Jonathan Alexander, "Labeur and Paresse: Ideological Representations of Medieval Peasant Labor," *The Art Bulletin* 72, no. 3 (1990): 436–52. The labors were prevalent in Christian architecture throughout Europe and, in the later Middle Ages, the Islamic world. See D. S. Rice, "The Seasons and the Labors of the Months in Islamic Art," *Ars Orientalis* 1 (1954): 1–39.
[157] "Psautier, avec calendrier," 13th century, Bibliothèque Louis Aragon, MS 124, Amiens, France.

and the Last Judgment (fig. 3.15, 3.16); here, the labors and zodiac—connected with small, sculpted sprouts of foliage—appear in a semi-circular arrangement, recalling earlier pictorial traditions that showed a deity, Jesus or sometimes Mithras, at the center of the image framed by these cosmological markers of time. In many ways, examples of the labors sculpted into the voussoirs of portals function similarly; At Chartres, where the tympanum above the north portal depicts an image of Christ encircled by angels and clouds, the labors and zodiac signs appear above the tympanum in two registers of alternating voussoirs in the north portal of the west façade (fig. 3.17).

Other examples display the labors of the months at a lower height, either beneath the feet of column figures or in the jambs themselves. These examples appear to have even more variation in the overall composition of the portal, especially the subject matter and main subjects depicted in the tympana or trumeau figures. At Saint-Denis, the signs of the zodiac and labors are split between two portals: the zodiac appears on the door jambs flanking the north portal of the west façade, while the corresponding labors appear on the south portal. The attributes, carved in roundels, are split into two registers on either side of each portal. At Senlis, small reliefs depicting the labors appear below column figures flanking a representation of the Coronation of the Virgin, which displayed in the tympanum (fig. 3.18, 3.19). At Sens, a similar ensemble of the Labors and the personified Liberal Arts appear in compact, rectangular reliefs on either side of the portal, which illustrates martyrdom of St. Etienne.

At Amiens, however, the reliefs are not only at eye-level, but enlarged and packed with dense detail (fig. 3.20, 3.21). These images are easily "read" and identified yet reward the viewer's close looking. One image that illustrates this particularly well is the 'warming' scene, in which a bearded figure warms himself by a fire during the winter months (fig. 3.22). This particular scene appears consistently in a number of image cycles, making it an ideal case study

for comparanda; a number of schematized examples survive from the twelfth and early thirteenth

centuries showing a single, seated figure facing a fire.[158] But in the Amiens relief, we have what

appears to be an elaborate interior scene: a bearded man wearing a heavy cloak has removed his

shoes—which he has placed under his bench in the bottom lobe of the relief—in order to warm

his feet by the fire, which he stokes with a poker while a cooking pot hangs above the flames. He

holds his other hand near the fire to warm his fingers and looks out towards the viewer. A piece

of furniture next to him even retains details of ironwork. At Amiens, the 'warming' scene appears

beneath the sign of Pisces.

Then as in now, the signs of the zodiac did not correspond directly to the twelve months

in the Middle Ages, but to the movement of the stars. A ninth-century manuscript conserved in

Munich, thought to have been copied at a scriptorium near Salzburg from a northern French

exemplar, not only illustrates each zodiac sign, the earliest surviving medieval zodiac images of

this kind, but also marks the signs of the zodiac in accordance with their overlap within the

twelve-month liturgical calendar. In a calendar matrix from the manuscript, the signs are shown

split over the twelve months, straddling months between their beginning and end (fig. 3.23). In

the detailed liturgical calendar, the movement of the sun from one sign to another is noted boldly

in red ink. Even though image cycles incorporated the signs of the zodiac, it should be

questioned whether each image in cycles like those at Amiens, Chartres, or Saint-Denis was

meant to correspond neatly to each month. Even more than their counterparts in books of hours

---

[158] See, for example, the archivolt at Chartres as well as the roundels from the Church of Saint-Lazare at
Avallon and the Cathedral of Saint-Lazare at Autun. The example from Saint-Denis appears to show two
figures.

and liturgical manuscripts, the representation of these attributes in the context of a monumental

portal evokes the passage of time through the change in seasons, rather than a precise calendar.

That said, when the labors are presented with the zodiac, there is variation within these

cycles as to which labor corresponds to each zodiac sign. In French image cycles, for example,

the 'warming scene' is frequently associated with Pisces and some, like the example in the

stained glass at Chartres, are even labeled with the month of February. But in other cycles, like

the roundels from Autun, the warming scene appears in conjunction with the sign of Aquarius,

and in the mosaic pavement at Otranto Cathedral, the warming scene is labeled with the month

of January. In the Creation Tapestry from Girona Cathedral in Spain, even though the

personification of February is fragmentary on its fraying edges, the tapestry shows the partial

torso of a man carrying two hunted birds. In this example, the indoor 'warming scene' actually

does not represent a month, but a personification of the season of winter, where it appears on the

top register of the tapestry with the other three seasons; the month of February, on the other

hand, is illustrated with an outdoor scene representing the hunt.

The variation in individual image cycles provokes questions about the experience of

climate in different geographic locations. The decision to place the 'warming scene' later in the

cycle at Amiens, for example, seems notable and might even suggest that a longer "indoor"

season was closer to the experience of local inhabitants, though it is difficult to say for certain

whether the quatrefoils, detailed as they are, reflect any medieval experiences accurately. Many

of the Amiens reliefs do reflect, however, an interest in showing representations of the

seasonality of plant growth more detailed than their counterparts. Given the other unusual

treatment of plant life in the sculpture and architectural decoration of Amiens, perhaps these

reliefs have more to reveal in regard to the understanding of the iconography of the north portal.

For example, the quatrefoil corresponding approximately to the month of July depicts a shirtless man wielding a sickle in bare feet, his pantaloons tied up around his waistband to expose his ankles (fig. 3.24). He stands on ground that appears to be covered with small flowers. The composition is similar to the example at Saint-Denis, but more dynamic; the figure holds his body at a diagonal, counterbalanced by the long sickle, the curving blade corresponding to the curvature of the quatrefoil's frame. Next to this relief, the quatrefoil representing Leo shows a similar shirtless figure gathering hay into bales, the finished bales stacked in the background on the left lobe of the frame (fig. 3.25). While there are certainly indications of class in these depictions of agricultural labor, the state of undress not only signals their identity as "rustics" or agricultural workers, but also indicates a practical response to the summer heat, removing a tunic or shoes to engage in physical activity. Contemporary manuscript representations of the subject rarely, if ever, depict agricultural workers bare-chested, while similar examples in monumental sculpture are also rare.[159]

Indications of the changing weather of the seasons can also be seen in the unique way plants are depicted in the reliefs, especially in the changes depicted from the spring months into the summer. In the oft-cited relief depicting vine cultivation, a fully clothed man digs in the soil between two vines (fig. 3.26), but notably, the vines do not have any leaves—this labor is being performed in the cooler days of early spring, before leaves have sprouted on the vines (it is worth recalling the many other vines depicted on the west façade that show naturalistic, three-point serrated leaves in conjunction with a vine). Above, the quatrefoil representing Aries depicts a ram surrounded by leafless trees (fig. 3.27). This set of quatrefoils makes an intriguing

---

[159] The scything image at Saint-Denis appears to show a bare-chested worker, but the accompanying reaping image in that series depicts the worker fully clothed. A French Psalter-Hours from c.1270 includes a roundel of a bare-chested man with a scythe, but the majority of manuscript illustrations show the figure clothed. See Psalter-Hours, Philadelphia Free Library, Ms. Widener 9, fol. 15v, reproduced in Hourihane, *Time,* 128.

comparison with the adjoining pair representing the sign of Taurus. Below, we see a falconing scene, in which the falcon lowers its beak to eat a morsel from a man's hand. The figure is flanked by leafy trees on either side, placing them in a tree-covered wood; above, the Taurian bull is surrounded by budding trees. The left lobe of the Taurus relief is blank, either incomplete or damaged, but perhaps it also once contained a third leafy tree, making this image rhyme visually with its counterpart to the left. These quatrefoils show the gradual process of blossoming, bearing witness to the transformation of spring in local time. In other cycles, these springtime images are rarely as specific in depicting the onset of warmer months and blossoming of plants.

The spring images culminate in the most foliate of the series, a pair of quatrefoils representing the sign of Gemini. While these images contain subjects that refer to traditional precedents, the Amiens versions are, again, intriguingly distinct from other contemporary depictions. In other images of Gemini from the first half of the thirteenth century, an artist might represent the sign as a Janus-faced man or twins that appear as mirror images of one another. At Amiens, however, the Gemini twins stand flanked by foliage; one figure wears a long gown and the other a shorter garment, one with shorn hair and the other with longer hair. The figures wear matching amulets and hold hands; the long-gowned figure holds a hand to the abdomen. This version of Gemini combines the zodiac sign with the image conventions of medieval courtship, a convention that is more commonly found in depictions from the later Middle Ages.[160]

Below the depiction of Gemini, the image of a bearded man in a tunic and cloak is seated in a verdant setting, holding a leafy branch in his left hand (fig. 3.28). The leafy bush to his left

---

[160] This image convention is more common in examples from the last decades of the thirteenth century or later, such as the representation of Gemini in the French Psalter-Hours from the Pierpont Morgan Library, Ms. M.729, fol. 10r, 1275-1299. See also Pierpont Morgan Library Ms. M.261, fol. 5b, 1495-1503; Pierpont Morgan Library Ms. M. 1003, fol. 5r, c.1465.

appears to shelter him, while a tree to his right provides a perch for a bird. This composition

recalls not only the blossoming of Aaron's rod in the south portal, but the sheltering foliage

beneath the column figures of Nahum, Micah, and Jonah. The relief is a composite of biblical

prototypes within the sculpted calendar, projecting biblical precedent into the viewer's reality

where he is both sheltered and chosen. It offers a pictorial connection between the image cycle of

the labors and the illustration of biblical texts. In comparison to other image cycles, this image is,

yet again, unusual; falconry is often depicted in association with Gemini, sometimes with a

mounted rider, as at Saint-Denis. Images of a figure holding two plants are often associated with

Taurus, like the stained-glass example at Chartres. The example at Amiens is unique because the

figure is seated and sheltered beneath boughs of blooming foliage, at ease in nature among the

burgeoning plants. Like the image of Gemini, later iterations of this image type appear to show a

man enjoying nature in a garden, but comparable images contemporary to Amiens or before its

execution, to my knowledge, have not yet been located or studied.

If the vices and virtues frame the Last Judgment, while biblical scenes frame the life of

the Virgin Mary, it is left to consider why the labors of the months and signs the zodiac were

selected to frame the portal of Saint Firmin. While the labors and zodiac frame images of Christ

in a number of twelfth- and thirteenth-century contexts, at Amiens, they frame the trumeau figure

of Firmin and the tympanum above, the depiction of the Invention, the miraculous warmth that

prompted visitors to come from neighboring cities, and the trees that burst into bloom in the

middle of January. It is significant that these quatrefoils, which outline the typical expectations of

climate and agricultural production, frame the first known illustration of this environmental

miracle, one that subverts meteorological and seasonal expectations completely. This

juxtaposition recalls Peter Abelard's discussion of miracles in his commentary on the Genesis creation narrative, in which miracles are events that are "against or above nature" (*contra vel supra natura*), a miracle as a subversion of order.[161] The portal, then, visually reinforces the miraculous events around the relics of Saint Firmin, advertising the authenticity and efficacy of their power.

Furthermore, the order of the zodiac and labors at Amiens is also unusual. Typically, zodiac cycles "begin" with Aquarius, corresponding roughly to January, or with Aries, if the calendar is oriented to the start of the astrological new year. The zodiac reliefs at Autun, Saint-Denis, Notre-Dame of Paris, and Strasbourg, for example, begin with Aquarius and end with Capricorn, while Chartres has an unusual arrangement that begins with Aries, starting at the left innermost voussoir. The Amiens series, curiously, follows neither of these formulae. Reading the series from left to right from the left side of the portal, the Amiens calendar appears to "begin" with Cancer. Curiouser still, the right side of the portal "begins" with Capricorn. Notably, the trumeau and the tympanum appear between the winter quatrefoil pairs representing Sagittarius/winter sowing to the left and Capricorn/feast preparations to the right. At Amiens, it appears that the order of the zodiac has been *rotated* to situate these winter months at the portal's center. The traditional start to the year is not marked spatially in this particular ensemble; rather, the pictured miracle in the tympanum interrupts the space, time, and meteorological expectations of the sculpted calendar itself.

---

[161] Qtd. In Benedicta Ward, *Miracles and the Medieval Mind: Theory, Record and Event 1000-1215* (Aldershot: Scolar, 1987), 5.

3.4 North Portal Tympanum: The Invention of the Relics of Saint Firmin

Turning to the tympanum, the miracle of the Invention appears in two of three separate registers. The bottom register echoes the seated patriarchs in the south portal, featuring six bishops seated on micro-architectural thrones, signaling the connection between the bishop and physical body of the church. Between them, the trumeau of Saint Firmin is surmounted by a grand daïs, the top of which resembles a châsse, undoubtedly a reference to the physical shrine that holds the saints' relics within the cathedral. Above, the central register illustrates the discovery of the body of Saint Firmin by Bishop Salvius, who wields a shovel and holds his hand up in a blessing. He is surrounded by youths, also with shovels, as well as men and women holding their hands in prayer (fig. 3.29). The body of Saint Firmin is portrayed at a diagonal, as if he is in the process of being uplifted. Saint Salvius' blessing hand also points to an obtuse twisted rod, presumably representing the miraculous light that illuminated the location of Firmin's burial, and this sculpted beam of light stretches up to the upper register, where an undulating, cloud-like texture occupies the sky.[162] This texture frames much of the image, recalling images and reliefs that depict a heavenly realm, such as the north tympanum from the west façade of Chartres.[163]

Unlike the example at Chartres, however, these sculpted clouds not only signify a heavenly realm, but the sweet-smelling odor and miraculous heat from the miracle story. According to the textual account, it was the odor that attracted the clergyman and laypeople from

---

[162] "Et sepultus fuit in Ambianensium civitate: locus tamen, ubi requiesceret, ignorabatur. Tempore autem S. Salvii episcopi corpus ejusdem Martyris repertum est. Nam, cum idem episcopus una cum clero & populo triduo ad Deum exorasset, ut eis corpus S. Firmini martyris revelaret, nimio odore fragrante, & lumine celitus super locum effuso, defossa ibidem humo, corpus sacrum inventum est." *AASS,* Sept. VII, 35.

[163] For another detailed reading of depictions of ephemeral textures in medieval stone sculpture, see Conrad Rudolph, "Macro/Microcosm at Vézelay: The Narthex Portal and Non-elite Spirituality," *Speculum* 96 no. 3 (July 2021): 601-663.

the neighboring towns of Cambrai, Noyon, Beauvais, and Thérouanne.[164] The inhabitants and

architecture of these four towns flank the central scene of the discovery of the tomb, two towns

on either side; the aromatic clouds make physical contact with the four micro-architectural

groups, touching each surrounding city. This specificity is significant, as this textured field also

comes into contact with the central scene of the discovery of Saint Firmin's tomb, suggesting that

this is the depiction of the miraculous odor which compels the inhabitants of the neighboring

towns to march towards Amiens, to sound their trumpets and bells (fig. 3.30).

The top register of the tympanum depicts the procession back into Amiens. It is at this

point in the text that the heat emanating from the body of Saint Firmin compels the trees to burst

into bloom as the body is carried in procession back into the city of Amiens. In this register, leafy

trees can be seen on the far right behind the representation of the *homme vert* wearing a foliate

crown and bearing a foliate branch.[165] On the far right, youths climb the blooming trees to watch

the procession, a reference to Christ's triumphal entry into Jerusalem. The cloud texture

continues along the outer edges of the uppermost register and flows to the apex of the

tympanum, where it is met with a group of angels waving censors and celebrating with music,

connecting the miraculous odor with liturgical ceremony, earthly music with heavenly music.

The proximity of the physical representation of the odor and the leafy branches to the far left of

the register suggests that in addition to the miraculous fragrance, this texture also represents the

---

[164] "Putabant omnes, qui astabant, intra sepulcrum tanti Martyris candida lilia & vernantes rosas, aliosque virentes herbarum flores pulchre pullulare: & quanto plus ad sacrum mysterium accederent, tanto amplius immensus odor flagrabat , & totam diœcesim urbis Ambianensium irrigabat, & usque ad omnes comprovinciales civitates late spargebat amœnitatem largiflui odoris. Omnis quidem multitudo urbis Taroanensium, Cameracensium, Noviomensium & Belvacensium, repleta fuit in ipsa hora suavitate & delectatione ipsius amœnissimi odoris sic, quasi in paradyso delitiarum se adesse putarent." *AASS*, Sept. VII, 34.
[165] Stephen Murray, *Notre-Dame, Cathedral of Amiens*, 115.

miraculous heat. In the tympanum, three invisible phenomena—heat, odor, and the heavenly

realm—are combined with one visual device.

The top register connects the châsse in the daïs above the trumeau with the illustration of

its ceremonial use. Two processing groups are shown in the center: the first group bears liturgical

objects—a cross can be seen behind the shoulder of the clergyman who carries the sacred text—

while the second group bears a medieval châsse. The carved image of the châsse used in this

register shows, remarkably, six registers separated by microarchitecture, a common form for

medieval shrine reliquaries of the twelfth and thirteenth century. It also recalls, however,

descriptions of the now-destroyed, thirteenth-century reliquary of Saint Firmin, presented on

October 16, 1204, commissioned before construction on the cathedral began.[166] The new châsse

had featured six enamel registers on either side illustrating the life and invention of the relics of

Saint Firmin; this stone version wrought in miniature appears, remarkably, to reference the actual

object.[167] This procession is taking place, it appears, not right after the miracle of the discovery

of Saint Firmin's body, but in the present. While the discovery of the relics of Saint Firmin

belongs to the past, the miracle of his relics belongs to the continuous present.[168]

---

[166] Corblet, 120.

[167] "1) S. Honorat donne à Firmin le bâton pastoral; 2) S. Firmin fait ses adieux à sa famille et à S. Honeste; 3) il convertit Arcade et Romule; 4) il est chassé de Beauvais par Sergius; 5) mort de Sergius, jeté à bas de son cheval; 6) S. Firmin érige à Beauvais une église, sous le vocable de Saint-Etienne; 7) conversion de Faustinian; 8) Firmin guérit un aveugle et deux lépreux, près de la porte Clypéenne; 9) Sébastien ordonne la décapitation de S. Firmin; 10) S. Salve découvre miraculeusement ses reliques; 11) leur translation; 12) les habitants de Beauvais, de Noyon, de Térouanne et de Cambrai vienne y assister." Corblet, 120.

[168] Gaposchkin, Cecilia. "Portals, Processions, Pilgrimage, and Piety: Saints Firmin and Honoré at Amiens." In *Art and Architecture of Late Medieval Pilgrimage in Northern Europe and England*, edited by Sarah Blick, 219.

3.5 Interpretation: Authenticity and Miraculous Weather

The meteorological event of the Invention is clearly out of step with the expected progression of the seasons and expected annual agricultural activity. Furthermore, the representation of Saint Firmin on the west façade and the depiction of the Invention of the saints' relics in the context of copious foliate sculpture and images of differing weather conditions show a certain concern for the authenticity of the local saints' miracles, a visual refutation to controversies of the twelfth century and an accompaniment to the saint's promotion, as well as the reinforcement of local liturgical activities, such as the procession of the relics of Saint Firmin and the liturgical ceremony of the *homme vert*. The inclusion of the labors of the months finds a final connection with the relic controversy at Saint Denis: at the chapel dedicated to Saint Firmin at Suger's abbey church, presumably where the attested Firmin relics could have been kept, a mosaic pavement dating to the twelfth century also contained the image cycle of the labors of the months. Today, the pavement survives in fragments and a drawing from the late eighteenth century, which depicts a tonsured figure in a monk's robe named Alberic surrounded by the four elements and medallions of the labors, likely commissioned after Suger's tenure.[169] The emphasis on the labors at Amiens makes an intriguing comparison, and might even represent an attempt to claim ownership of the saint through visual means.

Through the comparisons with the central portal and Christological details like the tree-climbing youths, the portal presents an image of Firmin as a local Christ figure. Just as Christ

---

[169] See Xavier Barral y Altet, "The Mosaic Pavement of the Saint Firmin Chapel at Saint-Denis: Alberic and Suger," in *Abbot Suger and Saint-Denis: A Symposium*, ed. Paula Lieber Gerson (New York: Metropolitan Museum of Art, 1986), 245–56. Barral y Altet points to evidence that the mosaic was likely commissioned after Abbot Suger's tenure, though Suger did commission a mosaic for the north tympanum of the west façade. The labors frequently appeared in floor mosaics in Italian architectural contexts, an example of which can be found in the twelfth-century floor mosaic in the crypt of San Savino in Piacenza. See Charles E. Nicklies, "Cosmology and the Labors of the Months at Piacenza: The Crypt Mosaic at San Savino," *Gesta* 34, no. 2 (1995): 108–25.

appears in three forms in the central portal, as the Beau Dieu, overseeing the Last Judgment, and at the apex of the tympanum as the apocalyptic Christ, Firmin is also pictured three times in the north portal: first as the trumeau figure of the bishop, then his deceased but preserved corporal body in the open tomb, and finally his relics paraded to the cathedral in the diocesan châsse. These three representations correspond to three temporalities: the life of the saint, the discovery of his tomb, and the ongoing liturgical processions of the cathedral. The flowering of trees in the miracle of the Invention also echoes the documentation of the environmental miracles that ensued after periods of harsh drought; by the intervention of saints, the heavens opened up and allowed the agricultural activities of the diocese to prosper, turning barren fields into bountiful harvests.

In sum, sculpted foliage ties together several iconographic themes across the west façade of Amiens: Marian, Christological, proto-apocalyptic, as well as themes relating to local lore and hagiography. While several foliate themes were chosen to illustrate scenes related to the prophets, the reliefs of the "wise and foolish" trees in the central portal point to foliage as a sign of grace which follows from saintly behavior in the earthly realm. The labors of the months in tandem with the representation of Saint Firmin show foliage and unexpected temperate weather as a sign of the miraculous at work. This meteorological transformation is presented in the context of sacred space sanctioned by the seat of episcopal power. What follows is an entanglement of natural phenomena—including weather and the regeneration of plant life—with religious belief and sacred space. This entanglement is not only represented visually on the west façade of Amiens, but in practice. The cathedral's sculpture presents an ordering of time and expectation of meteorological events, while providing ample parables for what happens when men and women turn away from the Church: plants wither and die, fields dry up and become unproductive, cities fall like ripe figs from a tree. Leafy and leafless metaphors become an

efficient vehicle for making resonant images to communicate this message. At Amiens, its

entanglement with broader iconographic narratives is not accidental, but deftly accomplished.

4

The Foliate Frieze at Notre-Dame of Amiens

Eugène Emanuel Viollet-le-Duc's drawings of foliate sculpture from Notre-Dame of

Amiens show a remarkable attention to detail and sensitivity. Having taken over the restoration

of the building in 1849, Violletß-le-Duc worked closely with the Duthoit brothers, local

Amiénois sculptors, to replace and in some cases completely redesign hundreds of individual

pieces of foliate sculpture throughout the exterior and interior of the building.[170] One element in

particular, however, was never fully understood by the architect and expert draftsman: the robust

foliate frieze that runs below the triforium the entire length of the interior. In a small drawing that

once belonged to the Viollet-le-Duc familial archives now at the *Mediathèque d'architecture et*

du patrimoine in Paris, he attempts to depict a small segment of the foliate frieze that travels over

a grouping of engaged shafts (fig. 4.1, 4.2).[171] In the drawing, Viollet-le-Duc not only captures

the foliage traveling over bay divisions, but also a small segment of the straight frieze continuing

beyond this example, labeled "A." This graphite sketch was the basis for the engraving that

---

[170] Documentation of the early stages of the restoration and Viollet-le-Duc's work with the Dutthoit brothers
can be found in Michel Barjon, ed., "La Cathédrale d'Amiens Façade Occidentale: Les Restaurations des XIX
et XX siècles" (Direction régionale des affaires culturelles de Picardie, Conservation régionale des monuments
historiques, Groupe de Recherche Art Histoire Architecture et Littérature, Paris, 1993), Bibliothèque Louis
Aragon, Amiens, France. See also *Aimé et Louis Duthoit: derniers imagiers du Moyen Âge* (Amiens: Musée de
Picardie, 2003); Raphaële Delas, "Les frères Duthoit, « derniers imagiers du Moyen Âge » ?," *Revue d'art
contemporain*, no. 07 (June 15, 2008): 74–89; Jacques Foucart-Borville, "Viollet-le-Duc et la cathédrale
d'Amiens," *Bulletin de la Société des antiquaires de Picardie*, Tome LIX (1981-1980): 172–238.

[171] Graphite drawing, album 2012/024/31, Mediathèque de l'architecture et du patrimoine, Paris, France. Other
drawings by Viollet-le-Duc from the restoration of Notre-Dame of Amiens are kept in album 50963, as well as
the Archives départementales Somme V 431.104 and two albums on the cathedral's restoration (no. 4 and 5) in
the Duthoit family archives in Amiens.

illustrates the frieze in the *Dictionnaire de l'architecture française*; in the entry *"Bandeau,"* the

engraving, a mirror image of the graphite original, appears in concert with a cross-section of the

frieze's molding profile and indication of its approximate height, just shy of half a meter tall.

This image stands apart from the Amiens corpus of working drawings, however, because

of its lack of specificity. While Viollet-le-Duc was meticulous in recording surviving fragments

of the rose frieze on the west façade, foliate capitals, and exterior foliate bands, the leaf forms in

the drawing bear more resemblance to Viollet-le-Duc's work in the King's Gallery (fig. 4.3, 4.4)

than comparable passages in the frieze. While Viollet-le-Duc's drawing shows trilobed leaves

divided by an approximate horizontal axis, photographs and simplified drawings show that the

frieze is comprised of individual leaf forms that curve upwards or downwards, interspersed with

fruits and vine tendrils (fig. 4.5, 4.6). Isolating these forms to understand their complexity, one

can see how Viollet-le-Duc might have simplified the units of the frieze in his drawing, resulting

in jumbles of sculpted foliage that stray from the actual object in the published engraving. The

point of considering Viollet-le-Duc's illustration is not to draw attention to the architect's

shortcomings; rather, it is to say that the ingenuity of this frieze deceived even this master

draftsman who had intimate knowledge of the building and its many modes of sculpted

decoration. Even he could not fully comprehend the frieze's composition.

Viollet-le-Duc's apparent struggle with depicting the frieze at Amiens highlights a larger

issue in the study of this unmatched example of foliate sculpture—and foliate sculpture more

broadly. It is perhaps not surprising that art historians who have studied the interior foliate frieze

at Notre-Dame of Amiens have come to widely divergent conclusions about its form and

meaning. The historiographic narrative of the frieze is one of evasiveness: the frieze, at its great

height, evades comprehensibility, its forms not easily described or understood—a contrast to

Panofskian interpretations of Gothic architecture that favor comprehensibility and clarity. At a time when sculptors were notably capable of representing naturalistic plant forms—elements such as recognizable vine leaves and climbing wild roses with thorns, as seen on the west façade at Amiens—the carvers responsible for the frieze elected to use more abstract, tri-lobed leaf forms (fig. 4.7). In this case, a close look at the object of study is warranted to clarify issues of appearance and manufacture while placing the frieze within the context of other examples of foliate carving and foliate design. It will become clear that the unprecedented scale and variation in the foliate frieze at Amiens put it at odds with all earlier examples, representing an innovative rupture rather than a continuation of an enduring trend in architectural ornamentation in Gothic architecture. Furthermore, the implications of this frieze, manufactured at great expense and composed of discrete sections of unprecedented variety, must be considered in concert with the chronology of the building and theories of serialized production at Amiens.

4.1 The Amiens Foliate Frieze in Scholarship and Imagination

The interior foliate frieze at Amiens has drawn commentary from many art historians, though interpretations have varied to a surprising degree, resulting in analyses that are sometimes at odds with one another. [172] It is notable that in the *Dictionnaire raisonnée,* Viollet-le-Duc does

---

[172] Eugène-Emmanuel Viollet-le-Duc, "Bandeau," *Dictionnaire raisonné de l'architecture française du XIe Au XVIe siècle,* vol 2. (Paris: Ve. A. Morel, 1874), 103-111, esp. note 1 on 108-110; Georges Durand, *Monographie de l'église Notre-Dame, cathédrale d'Amiens,* vol. 1. (Paris: A. Picard et fils, 1901), pp. 30, 213, 233. Virgile Brandicourt, "La faune et la flore de la cathédrale d'Amiens: lecture faite à la séance publique du 21 décembre 1905, de la Société des Antiquaires de Picardie," *Bulletin de la Société des antiquaires de Picardie,* 1905, Bibliothèque Louis Aragon, Amiens, France, esp. p. 27; Lottlisa Behling, *Die Pflanzenwelt der Mittelalterlichen Kathedralen* (Köln: Böhlau Verlag, 1964); Denise Jalabert, *La flore sculptée des monuments du Moyen âge en France, recherches sur les origines de l'art français* (Paris: A. et J. Picard et Cie, 1965); Stephen Murray, *Notre-Dame, Cathedral of Amiens : The Power of Change in Gothic* (New York, NY, USA: Cambridge University Press, 1996), esp. pp. 8-9, 59, 115 and *Notre-Dame of Amiens: Life of the Gothic Cathedral* (New York, NY: Columbia University Press, 2021), 254; Camille, Michael. *Gothic Art: Glorious Visions,* 136. New York: Harry N. Abrams, 1996. Dany Sandron, *Amiens, la cathédrale* (Paris: Zodiaque, 2004), 52; Mailan S. Doquang, *The Lithic Garden: Nature and the Transformation of the Medieval Church*

not consider the frieze in his general overview of foliate forms, the oft-cited essay "*Flore,*" but in

"*Bandeau,*" a shorter, more focused entry that highlights earlier Romanesque examples.

Although the frieze is illustrated in the *Dictionnaire,* the placement of the Amiens frieze within

this minor essay presents it as a curious anomaly rather than a singular achievement. In the text,

Viollet-le-Duc observes that thirteenth-century churches tend to de-emphasize the horizontal

line, but that the architects at Amiens felt the need to highlight the great height of the triforium

by including the foliate band. In a footnote, however, Viollet-le-Duc references to the frieze's

mixed reception among critics:

> We have often heard praise or blame by competent people for the arrangement of the
> large *bandeau* of Amiens Cathedral. But the truth compels us to add that the praises were
> given by lovers of Gothic architecture at its peak and blame by enthusiasts of the new
> style. Because there was a contradiction between the tastes and judgments of both parties,
> we do not quite know what judgment to make ourselves. We will only say that the
> solution adopted at Amiens is clear, that it denotes a definite intention, that this nave
> interior seems to us to be the most beautiful specimen that we have in French thirteenth-
> century architecture, that we can hardly imagine the interior without this rich belt of
> vigorously reworked foliage [*cette riche ceinture de feuillages vigoureusement
> refouillés*], or whether it would add or detract from the ensemble; and taking it to be very
> beautiful, executed by artists who were as good connoisseurs as we are and more familiar
> with its grand effects, we can only approve the boldness of the architect of the nave of
> Amiens.[173]

Even though he commends the "boldness" of the Amiens architects, Viollet-le-Duc also suggests

that the frieze's horizontality was, at least at one time, a topic of debate. We are left to wonder

---

(New York, NY: Oxford University Press, 2018), 28-34, 55.

[173] "Nous avons entendu souvent louer ou blâmer par des personnes compétentes la disposition du grand
bandeau de la cathédrale d'Amiens. Mais la vérité nous force d'ajouter que les louanges étaient données par des
amateurs de l'architecture gothique à son apogée, et le blâme par des enthousiastes du style roman. Comme
dans l'un ou l'autre cas il y avait contradiction entre les goûts et les jugements de chacun, nous ne savons trop
quel jugement porter nous-même. Nous dirons seulement que le parti adopté à Amiesn est franc, qi'il dénote
une intention bien arrêtée; que cet intérieur de nef nous paraît être le plus beau spécimen que nous possédions
en France de l' architecture du XIIIe siècle; que nous nous rendons difficilement compte de l'effet que
produirait cet intérieur dépourvu de cette riche ceinture de feuillages vigoureusement refouillés, s'il y gagnerait
ou s'il y perdrait; et prenant la chose pour fort belle, exécutée par des artistes aussi bons connaisseurs que nous
et plus familiers avec les grands effets, nous ne pouvons qu'approuver cette hardiesse de l'architecte de la nef
d'Amiens." Viollet-le-Duc, "Bandeau," 108-110 (my translation).

which "competent," though unnamed, critics thought the frieze was out of step with Gothic

verticality—it is possible that this was a topic of discussion during the restoration of the

cathedral. Despite highlighting the frieze's beauty in this footnote, Viollet-le-Duc characterizes

the vegetal band as an archaic *"souvenir"* of Romanesque architecture in the main text of the

entry.[174] The restorer-architect drew attention to the frieze and its remarkable singularity in the

margins of his text but did not consider it to be an innovation; rather, he highlights how this

unique decoration suggests that the master mason wanted to break the vertical lines of the nave.

Even in this focused text, Viollet-le-Duc's positioning of aesthetic judgments of the frieze strikes

a chord of ambivalence: it is a holdover from the Romanesque past if you read the main body of

the entry, the frieze is part of the most beautiful interior ensemble that Gothic architecture has to

offer.

In reflecting broadly on the building's foliate decoration in his monograph published just

after 1900, Georges Durand divides all examples of sculpted foliage, interior and exterior, into

two groups. The first group, according to him, is a freeform version of stylized foliage that, while

not corresponding clearly to any plant species, has a sense of movement and plant-like presence.

The second type is not found as frequently but includes more delicate sculptures that are directly

copied from nature, such as the vine, rose, and lily at the foot of the Beau Dieu and other

examples from the cathedral's portal sculpture. He includes detailed analyses of sculptural

elements often left out of architectural narratives: descriptions of complex foliate capitals in the

triforium, the variety of carved leaves and fruits in the nave aisle keystones, the many horizontal

---

[174] "Évidemment, ici, le maître de l'oeuvre a voulu rompre les lignes verticales qui dominent dans cette nef,
dont la construction remont à 1230 environ. Il y avait là comme [108] un dernier souvenir de l'architecture
romane. Sans avoir une aussi grande importance, il arrive presque toujours que les bandeaux, dans les édifices
du commencement du XIIIe siècle, passent devant les faisceaux de colonnes et servent de bagues pour
maintenir leurs futs posés en [109] délit" Viollet-le-Duc, "Bandeau," Dictionnaire, vol. 2 108-109.

foliate bands that articulate the cathedral's exterior. Durand appears to follow criteria set by

Viollet-le-Duc in the *Dictionnaire* that would be later taken up by Denise Jalabert and others,

setting "naturalistic" foliage in opposition to less precise forms (or in Jalabert's parlance, who

also comments on the Amiens frieze, "*la flore géneralisée*"). Durand characterized the non-

naturalistic foliage as more architectural, monumental, and gigantesque. The interior frieze,

however, is not included in this analysis, but takes on additional roles as an indication of

methods of production, construction phases, and taste.

Durand's analyses and observations of the frieze, rendered with surprising depth, could

be seen as attempt to rescue the frieze from dismissals of archaic forms and debates about

verticality. It is clear that he has grappled with the frieze's confounding scale and variety,

illustrating a cross-section of its undulating relief over the course of one bay and hand-drawing

its winding forms in otherwise precise elevation drawings (fig. 4.8, 4.9, 4.28). His descriptions of

the frieze also indicate a deep engagement with the object and a desire to understand its

manufacture. He notes the changing forms of the cathedral throughout its interior, writing that

the workmanship in the nave is "so original, having been completed after installation [*après la

pose*]," that "it is obviously the work of one hand." In the choir, however, Durand posits that the

repeating motifs, completely transformed from their nave counterparts, have been manufactured

before installation [*avant la pose*].[175] Such a notable change in production and design is clearly

puzzling to Durand, who investigates the nave frieze further in his analysis of the nave:

> This band is quite remarkable ... Although the stonemason has almost always avoided
> passing a joint through a leaf or a fruit, the connections in the stems are everywhere too
> exact and too precise for one to admit that the sculpture has been made before

---

[175] "Nous verrons dans la description générale du monument que le cordon sculpté d'une facture si originale,
fait après la pose, et qui n'est dû évidemment qu'à une seule main, fait le tour de la nef et du transept et s'arrête
contre les gros piliers 17a et 18a, tandis qu'au choeur ce cordon est d'un tout autre dessin et sculpté avant la
pose." Durand, vol. 1, 30 (my translation).

installation. However, in bays 9, 11a, there is a cluster [*grappe*] cut in half by a joint and
which has no counterpart, but this is the only case, and it may only be accidental.[176]

Like Viollet-le-Duc, Durand has been deceived by the frieze's clever design, positing that this

sculptural object with so much variation must have been carved by a single hand after

installation. Even though he recognized the placement of the masonry joints in the frieze and

questioned the accidental truncation of a sculpted fruit, he proposes a veritable revolution in

terms of sculptural production between the nave and the chevet, the transition from *après la pose*

to *avant la pose*. It is notable the effect the frieze had on these observers, particularly the ways in

which this sculpture appears to have captured their attention *as well as* their imagination: Viollet-

le-Duc cannot imagine the nave's interior without the great presence of this monumental foliate

band; Durand marvels at its changing forms. We can only imagine what either author might have

been able to achieve with a telephoto lens and digital photography; indeed, without these tools, it

is nearly impossible to comprehend the interlocking forms of the frieze as they transform across

the entire length of the cathedral's interior.

In regards to the issue of verticality and Gothic architecture, Durand praises the

horizontality of the foliate frieze and uses its presence to justify the good sense of the Amiens

architects. He also expands on this theory to reflect on Gothic architecture and verticality more

broadly: "The horizontal line is necessary for all good architecture," he writes, "it was its

systematic abandonment that was one of the main causes of the decline of the Gothic style."[177]

Here, the frieze itself is not a problem or anachronism; rather, the absence of horizontal elements

---

[176] "Ce bandeau est fort remarquable ... Bien que le tailleur de pierres ait presque toujours évité de faire passer
un joint à travers une feuille ou un fruit, les raccords dans les tiges sont partout trop exacts et trop précis pour
que l'on puisse admettre que la scupture en ait été faite avant la pose. Cependant, à la travée 9, 11 a, il y a une
grappe coupée en deux par un joint et qui n'a pas de contre partie, mais c'est le seul cas, et il peut n'être que
fortuit." Durand, vol. 1, 232 (my translation).
[177] "La line horizontale est nécessaire à toute bonne architecture; c'est son abandon systématique qui a été une
des principales causes de la décadence du style gothique." Durand, vol. 1, 233 (my translation).

in later examples of Gothic architecture led to the unraveling of the Gothic aesthetic system—

even though earlier iterations of Gothic, such as the cathedrals of Soissons and Noyon, also

forego emphasized horizontal decoration in their interior elevations. Above all, it is notable that

analyses of this singular element of foliate sculpture skirt issues of *taste* and *sense* in Gothic

architecture, recalling debates about propriety, morality, and decorum rehearsed in Renaissance

criticism of the *goti* and *tedeschi.* The status of foliate sculpture in Gothic architecture history

and criticism captures, perhaps more incisively than other topics, the difficulties inherent in

defining and explaining the principal characteristics of this architecture system.

In the decades since the 1980s, commentary on the frieze has focused less on its specific

formal qualities and more on questions of meaning, interpretation, and liturgical performance. In

addition to observing the frieze's distinct sections, Stephen Murray was the first to combine a

reading of the frieze in concert with local liturgical practice, specifically in the liturgical

celebrations of the Invention of the Relics of Saint Firmin in January. The frieze acts as an

additional garland festooning the church, not unlike the foliate crowns bestowed upon canons for

this festive occasion, and, in Murray's analysis, the changing forms of the sculpted foliage also

play a greater role as a metaphor for the signs of change in the cathedral.[178] In the same year of

Murray's first monograph on Amiens, Michael Camille remarked on the festive nature of the

frieze, positing that this sculpted garland may be a sign that actual garlands woven from plants

were hung from the stones of churches for liturgical celebrations.[179] While there is no written

evidence to support this specific use of plant life in the cathedral, plants did play an important

role in formal liturgical celebrations such as Palm Sunday as well as a range of parallel liturgical

---

[178] Murray, *Notre-Dame,* (1996), 8-9, 59, 115.
[179] Camille, *Gothic Art,* 136.

activities, including the blessing of crops and fields.[180] In Bavaria, *palmbuschen* processions, in

which foliate bouquets paraded to the church for blessing and taken home to be used as

apotropaic devices, are still enacted on Palm Sunday today.[181] Blessings of crops appear in

liturgical texts as early as the seventh century: four such benedictions appear in a seventh-

century Old Gelasian Sacramentary intended for Frankish use, including one for fruit trees, two

for new fruit, and one specifically for apples.[182] The presence of the stone foliate frieze in the

interior of Notre-Dame of Amiens suggests the importance of placing plant life in the sacred

space of the cathedral on a permanent basis, an ongoing conversation between sacred space and

the natural world.

Most recently, Mailan Doquang has also remarked upon the festive quality of the frieze.

Though Doquang's project concerns more general aspects of foliate bands in church architecture,

the Amiens frieze serves as an important point of departure.[183] Doquang also writes that the

frieze has a "garland-like" quality, but challenges Murray's thesis about the celebration of Saint

Firmin, highlighting the lack of foliate friezes in the tympanum of the portal of Saint Firmin and

pointing to the vine friezes in the central portal above the Beau Dieu. Because of this relationship

---

[180] Tamsin Rowe, "'Bless, O Lord, This Fruit of the New Trees': Liturgy and Nature in England in the Central Middle Ages," *Studies in Church History* 46 (2010): 53–65. Liturgical texts attesting to these plant-life ceremonies, many of them descendants or variations of the Rome-based Gelasian liturgy, circulated throughout the Carolingian kingdom and England throughout the Middle Ages. Rowe also completed a dissertation on the subject, which I was unable to consult: Tamsin L. Rowe, "Blessings for Nature in the English Liturgy, c. 900-1200" (Ph.D., University of Exeter, 2010). Karen Jolly also explores the protective prayers over agricultural production in the tenth-century *Durham Collectar* in Karen Louise Jolly, "Prayers from the Field: Practical Protection and Demonic Defense in Anglo-Saxon England," *Traditio* 61 (2006): 95–147.

[181] *Palmbuschenbinden* and the *Palmbuschen* procession are popular tourist attractions in Bechrechtsgarden, Germany, local traditions that claim roots going back to the Middle Ages. The *Palmbuschen* are hung in houses all year, traditionally in the attic, then taken down the following year to be burned for Ash Wednesday. Reinhard Haller, "Palmbuschen," in *Das Brauchjahr im Bayerischen Wald: Schriftliche Umfragen im ersten Drittel des 20. Jahrhunderts : Historische Quellen zur Volkskunde einer Landschaft* (Grafenau: Morsak, 2014), 239-240.

[182] Rowe, "Bless, O Lord, This Fruit of the New Trees," 57, note 20.

[183] Doquang explores the multifaceted potential for meaning behind of foliate imagery especially in light of architectural ideas that spread through the crusades, an art historical lineage that includes mosaic rinceaux at the Dome of the Rock and earlier exemplars in Cluniac establishments in France.

between vines and Christological images, Doquang suggests that the interior frieze is meant to gesture towards the Eucharist as a continuation of the central portal sculpture in the interior of the building. Returning to questions of production and manufacture, however, Doquang states that the frieze adheres to a "self-replicating" pattern that repeats bay-by-bay with "ordered complexity and exact repetition," an incorrect description that does not match the frieze *in situ*.[184] It is surprising that even in recent scholarship, there has been no consensus as to the formal qualities of this important frieze, let alone its interpretive possibilities. It was for this reason that it seemed necessary to photograph the frieze in its entirety to provide clarity on issues of pattern and questions related to the use of templates, as well as a universal description of the object that accurately represents its composite parts *in situ*.[185] Attending to the frieze as a sculptural object leads to deeper questions about why this monumental foliate sculpture might have been integrated into the architectural ensemble at Amiens, setting the design of the cathedral's elevation apart from its earlier cousins at Noyon, Chartres, Laon, and Notre-Dame of Paris. While many recent studies on medieval art and nature are exquisitely rich in historiographic analysis, the objects themselves still have much to offer.

The notably different sections of the Amiens frieze have been invoked in discussions of building chronology. Durand used the frieze to define breaks in the construction sequence of the cathedral, suggesting—erroneously—that the cathedral had two main phases of construction, as determined by the differing sections of the frieze in the chevet and the rest of the building and

---

[184] Doquang, *The Lithic Garden*, 34.
[185] While Durand took care to represent the frieze in its non-replicating entirety in his elevation drawings, subsequent studies opted for a single photograph showing a small portion of the frieze, usually taken from an oblique angle on the ground level of the cathedral. This mode of representation has likely perpetuated the idea that the frieze repeats, that a small part can stand in for the whole.

their different modes of production.[186] Stephen Murray was the first to observe that, in fact, the

frieze appears to have at least *three* distinct sections: the nave and western transept walls, the

north/south transepts and eastern transept walls, and finally, the chevet.[187] Murray's chronology

of the building, however, is not focused on the frieze; he determines that Luzarches completed

the building up to the interior dado in the first five years, then was joined by Thomas de Cormont

to bring the southwest aisle by about 1230. Thomas then oversaw construction on the northwest

aisle, the transepts, and the lower choir up to the arcades in the following decades. The upper

choir, overseen by Thomas' son Renaud, took place in the mid-1260s. Given the three master

masons and the three sections of the frieze, it is tempting to assign each mason a section. This

chapter will provide space to question the relationship of the frieze's changing forms to the three

known masons as well as other forces might have been at play in its changing manufacture.

Above all, the diversity of analyses and reactions to the frieze highlights the confounding

nature of the object—and the cleverness of the medieval artisans who made it. As a sculptural

object, the frieze is caught in a liminal space between innovation and reiteration, referring to

older modes of architecture while signaling a locally specific mode of architectural production

that is decidedly new. This chapter is meant to provide a way forward for understanding not just

this example at Amiens, but other unusual instances of foliate sculpture that evade traditional

iconographic frameworks and tools of art historical analysis. Combining a new analysis of the

---

[186] That Durand did not note the change in the transepts is surprising, given that the detailed elevation drawings in his monograph appear to reflect this change. Whether this was a difference in opinion between the archivist-historian and engraver is hard to say.

[187] Murray, *Notre-Dame*, (1996), 8-9. Murray, *Notre-Dame of Amiens* (2021), 241-254. Remarkably, we know the names of the first three master masons of the cathedral's construction because of their inclusion a plaque installed in the cathedral's pavement: Robert de Luzarches, Thomas de Cormont, and Renaud de Cormont. Puzzling out the building's chronology in relation to the tenure of these three masons has been an important aspect of Amiens scholarship.

frieze with a near-complete photographic reproduction (Appendix A) allows the frieze and the cathedral to be seen anew.

My analysis confirms the three sections determined by Murray and adds that there is, in fact, further variation within them which is truly unprecedented in Gothic foliate decoration. Furthermore, irregularities in the southeastern transept wall as well as the frieze's apparent change in direction in the north transept and northeastern transept wall raise further questions about production and stylistic choice. In taking on these issues in concert with the current understanding of the building's chronology, the frieze provides an unusual opportunity to reflect on the relationship between these distinct phases of the frieze's production and the cathedral's three known master masons; in terms of the dramatic change in the chevet, it is still possible to see a revolution in sculptural articulation, but perhaps not the one Durand imagined. Instead, I consider this to an indication of an aesthetic revolution and re-ordering of decorative priorities, the reverberations of which can be felt in the later thirteenth century, especially in Paris at the Sainte-Chapelle. Pondering issues related to the manufacture of the frieze may not fully untangle complex issues but documenting its variation and dynamic change certainly adds to the existing material record and expands the field of inquiry into the building's decoration, liturgical use, and questions of meaning in architectural and sculptural form.

4.2 Foliate Friezes in France before 1220

Before analyzing the Amiens frieze, it is helpful to review the visual comparanda against which this frieze has previously been understood. Looking at the form and placement of these earlier friezes clarifies the unique characteristics of the Amiens example and opens questions into its relationship to the building as well as its manufacture. As mentioned above, art historians have often considered the frieze at Amiens to be a descendent of foliate bands and string courses that articulate elevations of twelfth-century Romanesque and Early Gothic architecture, especially those found in the south transept at the cathedral of Soissons and the chevet of Saint-Remi in Reims. Viollet-le-Duc, for example, placed the Saint-Remi and Soissons examples together in the *Dictionnaire,* even illustrating both friezes with the same engraving in the entry *"Bandeau,"* despite their apparent differences (fig. 4.10). This encouraged a number of subsequent scholars of medieval architecture to group these examples together; most recently, Dany Sandron referred to the Amiens frieze as an enriched version of these two potential sources. Doquang expressed doubt about the relationship between these friezes as well as the methodological utility of defining sources on a stylistic basis, given the diversity in French foliate friezes of the twelfth and thirteenth centuries. It is nonetheless useful to return to the friezes themselves if only to build a visual vocabulary for the types of forms one can expect to find in these sculptural embellishments, further defining the characteristics that make the Amiens frieze unusual.

At the cathedral of Soissons, the foliate frieze adorns the curving space of the south transept just below the gallery (fig. 4.11). Here, the frieze consists of mirroring rinceaux motifs sprouting round, five-point leaves with a fruit growing vertically between each motif. The variation in this frieze is minimal and, in comparison to Amiens, it is quite diminutive in scale and execution. The mirroring rinceaux pattern is a common twelfth-century pattern, appearing on exterior foliate friezes at Laon (fig. 4.13, 4.14).[188] At the church of Saint Remi in Reims, two distinct foliate friezes adorn the four-level elevation of the chevet: one frieze adorning the lower sill below the gallery, and another made of stout crockets that appears below the triforium (fig.

---

[188] A variation of this pattern also appears in the foliate bands between the registers of the central tympanum at Amiens, though much of the frieze has been damaged or restored.

4.12, 4.16, 4.17). The crocket frieze could be considered in concert with examples of exterior

crocket decoration such as Notre-Dame of Paris, where continuous friezes made up of crockets

appear on the exterior elevation of the cathedral, articulating the horizontal levels of the

elevation and outlining the geometry of rose windows and gables (fig. 4.18, 4.19). The lower

frieze at Saint-Remi is made up of non-mirroring rinceaux curling in the same direction, some of

which terminate in a leaf form or rosette. Some of the individual motifs are stretched to

accommodate the spacing of the bays and lose their spiraling, rinceaux-like appearance, but the

pattern is repeated continuously throughout the chevet.

This rosette pattern recalls examples at Autun, Beaune, and Vézelay, where friezes with

repeating inset rosette patterns appear in the interior elevations. At Autun, the frieze runs below

the stout triforium, while at Vézelay, a horizontal interior frieze divides the elevation in half

between the arcade and the clerestory above (figs. 4.20-4.24). These particular motifs are

resonant with antique foliate friezes found in the decorative borders of Roman sarcophagi, such

as the Mattei I sarcophagus, which were transmitted in a variety of media throughout the Middle

Ages, including ivory carving and architecture (fig. 4.24, 4.25). The decorative similarities

between sarcophagus, casket, and church resonate with themes of entombment, commemoration,

and the flowering of regeneration. As the analysis below will make clear, the frieze at Amiens

does not have mirroring rinceaux motifs nor does it have evenly spaced rosettes. Furthermore, it

is unusual for an interior frieze to match the scale and monumentality of the Amiens example;

before the construction of Amiens, foliate bands were typically limited to a discrete section of the

building or scaled down in size.

The frieze at Amiens is also distinctive from curling vine motifs and rinceaux derived

from antique sculpture, such as examples of foliate carving found in the cathedral of Saint-

Mammès at Langres (fig. 4.26). In the curved space of the chevet, spiraling rinceaux rendered in shallow relief alternate below the sill of the gallery, reminiscent of the circular, spiraling vines found on the lower register of the Ara Pacis. Masonry joints bisect the delicately rendered foliate forms, which repeat in regular intervals—the Amiens frieze, as outlined below, is constituted of self-contained blocks, each individually carved. Below the Langres rinceaux, another common type of acanthus-type motifs adorns the pointed arches, resembling similar friezes that appear in late antique carving and Carolingian manuscript illumination. This type of frieze can be found in several other twelfth-century buildings, most notably at the collegiate church of Notre-Dame-du-Fort at Etampes, where it frames the innovative sculpted ensemble in the south portal.

In terms of distinctive foliate bands, the heavily damaged portals of Noyon contain many extraordinary examples of foliate carving, well-preserved despite acts of iconoclasm following the French Revolution and German bombardments of the twentieth century (figs. 0.1, 0.2). The carvers at Noyon employed distinct foliate designs between the portals' voussoirs as well as a deeply undercut, horizontal band at the springing of the pointed arch in the north and south portal, an approach to foliate carving that was also adopted later in the west portals of Notre-Dame of Reims. The forms depicted in these friezes appear to represent specific species of plants, with naturalistic leaf forms resembling ivy and grape leaves, a departure from repeating rosettes and rinceaux motifs. These portals, which do not have a firm date but were likely in place before 1231, also share decorative similarities with those at Amiens, particularly in the floral diapering and integration of a single row of quatrefoils (now scraped bare).[189]

---

[189] Charles Seymour, *Notre-Dame of Noyon in the Twelfth Century; a Study in the Early Development of Gothic Architecture.* (New York: Norton, 1968), 66; Robert Branner, *Burgundian Gothic Architecture*, Studies in Architecture 3 (London: A. Zwemmer, 1960), 53, 86; Arnaud Timbert, and Philippe Montier. *La cathédrale Notre-Dame de Noyon et son quartier.* Noyon: Cap régions, 2012. See also Paul Crossley, ed. *Gothic Architecture*, note 41. Seymour notes documentation of a bell rung from one tower in 1231 at Noyon, signaling

While there is a wealth of examples to discuss, it is also important to note that scores of twelfth- and thirteenth-century churches do not include foliate bands into their decorative schemes. While foliate carving adorns the portals at Noyon, the interior includes no foliate decoration outside of the carved crocket capitals. The same is true at Laon, Notre-Dame of Paris, and Notre-Dame of Reims, where interior foliate elements are reserved for capitals. Because of the absence of interior foliate friezes in early Gothic buildings, the decision to include such a robust foliate element at Amiens is unusual. Given the occasional presence of these friezes in earlier church designs such as the Soissons transept, the Saint-Remi chevet, Autun, Saint-Mammès, and the nave at Vézelay, scholars seeking to understand the frieze have placed it in conversation with these earlier examples. But as we will see in the next section, the Amiens example stands alone in many respects: this frieze does not follow a repeating pattern, borrow motifs from common sources, or contain any true rinceaux or flower motifs; instead, occasional vine curls vary widely and appear sporadically and its unique, organic qualities make its forms appear to grow across the interior of the cathedral.

4.3 Description and Analysis: The Frieze in Three Parts

Notre-Dame of Amiens has many foliate bands that range in scale and complexity. Robust foliage articulates the exterior of the cathedral on its two towers, the west rose, and roofline, while a delicate rose frieze unites the three western portals. The central tympanum features two friezes of deeply undercut vines that separate the three registers illustrating the Last Judgment; these delicate vines follow at least four different patterns, some reminiscent of foliate

that the portals were likely complete by this date. The start date of the portals is not settled, though most scholars place the beginning of this construction phase between 1205-1215.

friezes found on the exteriors of Notre-Dame of Paris and Laon, while others appear to have

been replaced. Still another mode of rich foliage can be found on the Saint-Honoré portal,

framed by carved leaves and fruits that are densely lush and appear to have been carved directly

into the stone *après la pose,* perhaps in an effort to reconcile the dense iconography of the

sainted bishop Honoré illustrated in the portal's tympanum with the later addition of the *Vierge

Dorée* trumeau.[190] In the nineteenth century Viollet-le-Duc designed and executed a foliate frieze

crowning the King's Gallery as part of larger restoration efforts, its full, fleshy leaves punctuated

with foliate masks. The wild rose frieze that unites the three west portals, also heavily restored

by Viollet-le-Duc, was a unique inclusion to the sculptural ensemble, also adopted later in the

thirteenth century in the sculpted decoration of the Porte Rouge at Notre-Dame of Paris.[191] Many

of these friezes merit close study, but the interior frieze that runs along the bottom of the

triforium is not only unique to the program of vegetal sculpture at Amiens, but to Gothic

architecture writ large.

A visual orientation to the frieze is crucial to understanding its unique characteristics,

especially because it has never been the focus of a dedicated study or full formal analysis.

Compared to earlier string courses and foliate bands, the size of the frieze has been expanded to

suit the larger scale of the architectural ensemble at Amiens, dividing the elevation in half and

traveling over the bundled shafts of each bay. Its high relief produces a heightened contrast of

---

[190] John Ruskin, *Our Fathers Have Told Us: The Bible of Amiens, Chapter IV, Interpretations*, 4th ed. (London: George Allen & Sons, 1909), 148; Cynthia Gamble, "L'ombre de la vierge dorée de la cathédrale d'Amiens: Ruskin et l'imaginaire proustien." *Gazette des Beaux-Arts* 1995, 313-322. Ruskin thought the south portal's frame was decorating with hawthorn leaves, leading Proust to identify them in this way. This exchange highlights the power of suggestion in viewing and identifying foliate forms, especially those associated with the Virgin.

[191] Ruskin, *Bible of Amiens,* 166-167. Ruskin is an ideal witness for Viollet-le-Duc's restoration of the rose frieze. He notes that "only a few flowers have been spared here and there," and then elaborates in a footnote: "Compare its roses with the new ones running round the arches above—and you will know what 'Restoration' means." On the the Porte Rouge, see M. Cecilia Gaposchkin, "The King of France and the Queen of Heaven: The Iconography of the Porte Rouge of Notre-Dame of Paris," *Gesta* 39, no. 1 (2000): 58–72.

light and shadow, creating an emphatic divider between the flatness of the arcade wall below and the perceptible depth of the triforium above, between the monumentality of the arcade and the feathery lightness of the upper elevation, which is set back slightly. From its height of approximately 20.8 meters, the frieze appears to resist predictability; its forms do not have the look of mechanic repetition like those at Soissons or Reims but instead undulate organically. Viewing the frieze in profile from the pavement of the nave, one can observe that the leaves are carved in varying levels of relief, with some sections directed more obliquely towards the viewer on the ground and others aimed perpendicularly towards the other side of the nave (fig. 4.28). From this angle, it becomes clear that its forms do not adhere to a consistent pattern like earlier examples—it is no wonder that Durand imagined the frieze might have been carved *in situ* by a single hand. Furthermore, the majority of the frieze is well-preserved, with the exception of a few sections in the transepts and behind the organ on the west wall of the cathedral. Photographs from the removal of the organ in the twentieth century reveal that this portion of the frieze is in poor condition (fig. 4.29). The southeast transept wall is also heavily damaged, as is the southwest transept wall section adjacent to the southwest crossing pier. The frieze contains only one figurative element: a beast's head caught in the act of devouring one of the luscious fruits, found on the north transept (fig. 4.30).

While the frieze contains many variations, we can discern three distinct sections (fig. 4.31). Located in the nave and west transept walls, the longest and perhaps most prominent section of the frieze is characterized by vine curls, fruits, and tri-lobed leaf forms that terminate in circular, clover-like ends that curve gently upwards or downwards (fig. 4.32-A). The motifs appear in a random sequence; for example, vine curls are more prominent on the north nave wall and northwest transept arm than its southern counterparts, rather than regular intervals. The fruits and leaves also vary in size, execution, and in their relationship to the negative space behind the carving. In this portion of the frieze, the upper and lower torus are often visible (fig. 4.33). While certain leaf elements resemble one another, this resemblance does not produce the same rhythm of mirrored rinceaux or rosette friezes. In fact, looking for matches between disparate parts of the frieze only seems to emphasize the differences. These differences are especially apparent in bays 5, 6, and 7 of the south nave wall, where the scale and shape of leaf forms and fruits vary to an astonishing degree (see Appendix A). The southwest transept arm, though heavily damaged, has similar variations: in the aisle bay adjacent to the crossing, the intact leaves are smaller, respecting the invisible boundaries of the upper and lower torus. In the middle bay, these smaller leaf forms give over to expansive tri-lobed carvings and fruits that reach beyond the torus, fleshy leaf forms that fill more of the frieze's negative space. The north nave wall displays more consistency in terms of the size of the frieze's individual parts, while the northwest transept wall has more variation in terms of the size and execution of the carved fruits.

While the north nave wall is executed with greater consistency, we still cannot say that these motifs follow a distinct pattern, a point that will be explored in more detail below. Furthermore, this section of the frieze incorporates five unfinished sections: in bays 1, 3, and 4, the passage that travels over the bundled shafts has been mostly carved except for the fruits, which remain round and undefined. Two individual blocks in bay 6 have also been installed in an unfinished state, its vine curl and leaf roughed out, still clinging to the negative space that is undercut in neighboring units (fig. 4.34). Its fruit is round and missing the suggestion of berries. The incorporation of these unfinished segments raises a number of questions about the frieze's production and the relationship between the sculptors and the building's chronology. First, it appears that the fruits might have been carved last; was this because multiple artisans worked together to execute the individual blocks of the frieze, executing leaves and fruits separately? Or was this last step left unfinished because of time or labor constraints? It is also unclear whether this carving work lagged behind installation at this stage of the cathedral construction. Was the masonry of the north nave wall laid faster than sculptors could produce these frieze sections, or was there a shortage of funding or labor? While it is not possible to know why the unfinished portions of the frieze are concentrated in this area, another possibility may lay in the transition between the first two master masons at Amiens. According to Murray's chronology, the southwest aisle was built up to the level of the triforium first, work that concluded around 1230 under the direction of Robert of Luzarches and close collaboration of Thomas de Cormont.[192] The north aisle, however, was built after Thomas had taken over the cathedral's construction as master mason around 1235; perhaps the unfinished blocks were incorporated at this time because they had already been produced, but the transition, turnover of craftsmen, and loss of time or resources left Thomas little choice but to incorporate the unfinished sections.

The second section of the frieze occupies the transept façade walls and eastern transept arms. This portion contains several variations and irregularities but is generally characterized by numerous, ribbed vine curls, close-textured or conical fruits, and leaf forms that abandon the flat, tri-lobed forms of the nave for fluttering, sculptural effects (fig. 4.32-B and fig. 4.35). Notably, while the first section in the nave reads right-to-left, or counterclockwise, the second section now "grows" in the direction of the east end; in the south transept, the direction continues to be counterclockwise, but in the north transept, the leaves and fruits have their stems turned to the right, reversing direction. This reversal had to be pre-planned in the production of the individual units of the frieze; because the upper torus protrudes farther than the lower torus, this was not a

___

[192] See Murray, *Notre-Dame of Amiens,* (2021), 182-190, 241-254.

simple case of flipping the units horizontally; these sections would have to have been created deliberately to mirror the directionality of the motifs in the south transept. The effect makes the frieze appear to converge on either side of the chevet, growing towards the choir, creating a sculptural crescendo of movement in the crossing space.

The southeast transept arm is worth considering in more detail (Appendix A and Fig. 4.35-A). This part of the frieze differs subtly from the northeast wall and outer transept arms because of its straighter leaf forms with little or no curving axis, "walking" vine tendrils that reach delicately towards the outer edges of the frieze, frequent vine curls, and significant negative space. In the aisle adjacent to the choir, the section of the frieze adjoining the crossing pier incorporates two vertical leaf motifs resembling the pattern that constitutes the chevet, albeit less delicately rendered (Fig. 4.36). It is important to acknowledge this troubled portion of the building was also the site of the significant fire of 1258, which damaged much of the work in the eastern end of the cathedral, especially the southeast side. As Murray has illustrated, a significant portion of the stonework in this part of the building was refaced in the Middle Ages.[193] The stylistic difference and plodding execution of this section of the frieze suggests that this portion of the frieze was likely also part of the refacing re-facing efforts, including the insertion of the rough piece of the chevet pattern near the southeastern pier.

The chevet constitutes the third and final distinct section of the frieze: an outlier, this portion features palmette-like forms alternating with triangular motifs terminating in foliate crockets, many of which have now broken off (fig. 4.32-C). This portion of the frieze loses horizontal emphasis altogether; here, the alternating leaf forms appear to sprout vertically from the base of the frieze. Compared with the singular unit worked into the southeast transept wall,

---

[193] Murray, *Notre-Dame* (1996), 72-75.

the five-point leaves of the chevet pattern are more deeply undercut with fluttering dimensionality. Unlike the carved ivy, oak, and grape leaves in the chevet triforium capitals, however, these five-point leaves do not appear to reference a specific plant, even though they are carved with great sensitivity. Overall, the repeating forms in the chevet depart from the organic undulations of the first two sections of the frieze. The chevet section is mostly made up of units with two motifs each (and a few one-motif units as exceptions), the forms executed with greater uniformity than the first two sections of the frieze. Unlike its counterparts in the transept and the nave, this alternating pattern can also be found in other contexts in the latter half of the thirteenth century; notably, sculptors opted for a similar pattern in the foliate friezes circling the exterior nave buttress piers as well as the roofline of the radiating chapels at Amiens as well as the interior of the Sainte-Chapelle just below the level of the clerestory and its exterior, articulating the lower chapel, upper chapel and roofline (fig. 4.37, 4.38). The examples at Saint-Chapelle show tri-lobed leaf forms, though a portion of the frieze now on display in the lower chapel incorporates five-point leaves carved on a smaller scale (fig. 4.37-B). A similar frieze also runs along the interior of Notre-Dame of Reims, articulating the upper reaches of the outer wall (Fig. 4.39). Other examples of this type of alternating-leaf-and-crocket frieze share similarities with crocket friezes produced for a variety of contexts in the twelfth century, such as the ones found on the exteriors of Notre-Dame of Paris and Saint-Gervais-Saint-Protais in Soissons. Crocket friezes, which will be explored in the next chapter, experienced another resurgence in the later thirteenth century.

While narratives around Gothic architectural decoration often point to a foliate intensification over the course of the thirteenth century, the frieze at Amiens illustrates that this transition was more complicated than adding or enhancing foliate elements to the architectural ensemble. At Amiens, sculptors replaced the unbridled, directional, and organic forms of the nave with standard, repeating motifs in the chevet. Not only was this pattern more standardized, it appears in other buildings from the latter half of the thirteenth century in France. The sections in the nave and transept arms, one the other hand, were not replicated in another architectural context—or at least not one that survives. While sculptors might have subdued or "tamed" the foliate frieze in the chevet at Amiens, it should be noted that the foliate carving in the chevet triforium capitals is at its most exuberant just a few meters above, with inventive compositions and hypernaturalistic vines that travel diagonally across the capitals' sculptural field. Similar approaches to organic foliate design can also be seen in the dado capitals of the radiating chapels of the east end, where grape vines appear to grow organically (Fig. 4.40, 4.41). It is not so much that the sculptors were uninterested in foliate designs in this phase of construction, but that decorative priorities seem to have changed. The smaller chevet capitals of the dado and triforium became fields of hyper-realistic and inventive foliate sculpture, while the frieze became a site of standardization and verticality. We might also wonder again about the transition between master masons at this juncture, this time between Thomas de Cormont and his son, Renaud. While it is not possible to know definitively, the pattern could have changed to accommodate the aesthetic challenges of decorating the curved bays of the chevet, as several earlier examples of interior foliate friezes in France also adorn a curving, circular space: Saint-Remi, the Soissons transept, and Saint-Mammès at Langres. Additionally, since this portion of the building was enriched by a glazed triforium under the direction of Renaud (who also, it appears dealt with the fallout from the 1258 fire), the de-emphasis of the frieze might have also been a cost-saving measure to redirect funds towards these other decorative details. Murray, Mark, and Bork suggested that Renaud's openwork flyers were not only an aesthetic choice, but informed by strained economics: the reduced weight of the flyers would have required less scaffolding and centering, reducing the amount of timber and labor needed to complete the work.[194] The timber roof of Amiens, completed between 1280-1310, also makes use of a lightweight design requiring reduced loads of tall oak beams, suggesting an interest in building lighter—and more economically, with fewer materials needed.[195] Furthermore, Robert Fossier showed that records of timber yields indicated a steep decrease in Picardy in the last decades of the thirteenth century, though the reason for this is not clear.[196] Were resource shortages and price increases also driving design changes in the cathedral's construction, including the execution of sculptural decoration? There are many indications in the later construction phases that economic strain and shortages of timber, likely accompanied with rising timber prices, was an important factor in re-imagining the design of the choir and cathedral roof.

4.4 Variations in the Nave and West Transept Walls

Having defined the three sections, a closer look at the first two sections reveals that they are made up of dozens of individually sculpted blocks of varying widths, containing different combinations of deeply undercut leaf forms, fruits, and vine curls rendered in high relief. As art historians have noted, the overarching visual unity of the frieze is notable, despite these variations; consistently, the blocks that make up the frieze are the same height and, as noted in Viollet-le-Duc's engraving, maintain the same molding profile for the upper torus and lower torus. The leaf forms and fruits appear coherent from a distance, as if they are responding to the

---

[194] Robert Bork, Robert Mark, and Stephen Murray, "The Openwork Flying Buttresses of Amiens Cathedral: 'Postmodern Gothic' and the Limits of Structural Rationalism," *Journal of the Society of Architectural Historians* 56, no. 4 (1997): 285.
[195] Murray, "The Cathedral Is Crowned by a Roof and Spire," in *Notre-Dame of Amiens* (2021), 259-262.
[196] Fossier, *La terre et les hommes,* 311.

same lithic laws of nature. Within each individual block, the leaf stems begin at the same approximate latitude (though there are some exceptions), allowing for the sculpted plant to be read as a continuous garland or vine growing across the entirety of the interior, even as the direction, size, and execution of these elements change. It appears that these elements, at least, were tightly controlled in the frieze's production. If these elements were controlled during the production process, however, then why do the individual units vary to such an astonishing degree? Distilling the frieze into its distinct parts, it is clear that sculptors took liberties with the size, positioning, and style of the frieze's individual elements, playfully manipulating curling tendrils, the length and overall shape of the leaves, as well as the spacing of each element within each individual block.

Within these sections of the frieze, there is remarkable variety between the individually sculpted blocks that has not yet been acknowledged in art historical scholarship. The principal elements that make up the frieze include a variety of tri-lobed leaf forms, fruits, vine curls, and tendrils; each unit contains some combination of these elements. The masonry joints in the frieze rarely bisect a form; rather, nearly every block is self-contained and designed to work within a more coherent whole—the few exceptions to this rule seem to have been cut short in order to fill the length of a bay (see Appendix B). This suggests that each unit was executed individually in a workshop before being set into place. The only perceptible pattern is indicated by the treatment of the transitions at some bay divisions: a bundle of vine-like tangles seen at the right of bundled shafts in the nave and western transept arms—though, on closer examination, even these vine forms vary from bay to bay. These portions of the frieze were likely planned further in advance and benefitted from adherence to a general form; filling this curved space traveling over the bundled shafts was a more complicated endeavor than filling the space between the bays.

The variation is perhaps best seen in the south wall in the bays adjacent to the southwest

crossing pier. (Figs. 4.33, 4.42). This more experimental area contains several truncated leaf

forms that are executed in a manner that could be called clumsy; they stand apart from the longer

leaf forms that make up many of the units in the western bays and the north nave wall. While the

northwest transept wall is quite damaged, remnants of leaf forms suggest that these were also

carved in a diminutive mode similar to the south nave wall. This portion of the frieze is located

above the experimental pier capital noted by Murray as belonging to the early construction at

Amiens, in which sculptors appear to have changed their minds mid-manufacture, resulting in a

phantom torus surrounding its middle mimicking the capitals at Soissons or Laon (figs. 4.43,

4.44).[197] It follows that not all details had been thought out completely before the laying of these

first stones; it appears that sculptors faced a reckoning when determining the size and execution

of the foliate band. Furthermore, the variation and apparent iterative process of the sculpture runs

counter to theories about the serialization of stone cutting in the construction of the cathedral set

forth by Kimpel; in the case of the frieze, the process of its making was ongoing, subject to

continuous iteration as well as playful variation. Fortunately for this study, the inclusion of the

experimental forms in the walls of the cathedral record a breadth of human interaction with the

material and a clear artistic process that, like the individuality of its many sculptors, varied

organically.

While no two blocks are identical, there are several distinct block types that recur

throughout the length of the frieze (fig. 4.45). For the purposes of this study, these block types

have been labeled accordingly: single leaf, double leaf, single leaf with a vine curl, two

consecutive leaves followed by a vine curl, and triple leaf, which are rare. Three additional types

---

[197] Stephen Murray, "Looking for Robert de Luzarches: The Early Work at Amiens Cathedral," *Gesta* 29, no. 1 (1990), 121. Murray notes this unusual capital and its place in the earliest efforts of construction at Amiens.

appear in the eastern portion of the transept arms: alternating leaf-vine-leaf and vine-leaf-vine

blocks, as well as a single leaf form with two vine curls. Generally, the leaves appear to alternate

between an upward and a downward motion, but this pattern is interrupted several times

throughout the frieze and is not a universal standard applied to the whole object. The single-leaf

blocks and blocks with a single leaf and a vine curl appear in both upward and downward

variations, as do double-leaf block types. Looking at a double-leaf type, it appears at first glance

that the leaf pattern has simply been flipped; looking closer, however, the leaves are not mirror

images of each other but instead follow their own formal logic. This also challenges the theory

that sculptors used templates or serialization in the production of the frieze; it appears, in fact,

that the sculptors embraced or even aimed for a more organic "look" rather than clean

repetition.[198]

Considering this point further, it is notable that many variations exist within each type.

For example, there are dozens of two-leaf segments, but some are short while others are long,

requiring the leaves and fruits to be longer or shorter to fill the space; occasionally, extra

negative space has also been left blank (Fig. 4.46). Even within the length of one bay, there are

consecutive blocks belonging to the same type that are not the same width. This suggests that

sculptors might have worked with pre-cut stones in the fabrication of the frieze, making the leaf

forms fit the length of the stone block rather than cutting the block to the exact dimensions

needed for the frieze, as the sculptors appear to have done at Saint-Mammès. But even when

blocks are of similar length, type, and carving style, there is still remarkable variation in form; on

---

[198] Dieter Kimpel, "Le dévelopment de la taille en série dans l'architecture médiévale et son rôle dans l'histoire économique," *Bulletin monumental* CXXXV (1977): 195–222; Murray, "Looking for Robert de Luzarches," 118. Dieter Kimpel made serialization and standardization an important aspect of understanding the architectural production at, though Murray has pointed to changes in molding profiles to indicate that there was, in fact, more experimentation and change as the building was constructed. In regard to the frieze, the serialization theory was most recently put forth in Doquang, *Lithic Garden*, 34.

the northwest transept arm, for example, four consecutive leaf-vine segments illustrate this point

clearly: two of these examples have delicate vine tendrils extending downwards, one knotted into

a pretzel shape (4.47).

While some block types occur more frequently in particular sections of the frieze, there is

no pattern dictating the order in which the block types were placed (see Appendix B). While

certain motifs dominate in particular parts of the frieze—for example, vine curls are more

prevalent on the western end of the northern nave frieze and in the transept arms—there is no

discernible sequence that can predict the order in which the individual blocks appear. For

example, the westernmost bay on the south side of the nave contains nine distinct blocks, which

are laid in the following sequence:

Number	Compositional type
Block 1	Single leaf
Block 2	Leaf-vine
Block 3	Single leaf
Block 4	Single leaf
Block 5	Leaf-vine
Block 6	Double leaf
Block 7	Leaf-vine
Block 8	Leaf-vine
Block 9	Double leaf

Table 4.1: Color-coded compositional types of Amiens frieze, south nave wall, first bay from the west

The bay next to it, however, has just seven blocks, which follow a completely different sequence:

Number	Compositional type
Block 1	Double leaf
Block 2	Double leaf
Block 3	Leaf-vine
Block 4	Double leaf
Block 5	Single leaf
Block 6	Double leaf
Block 7	Double leaf

Table 4.2: Color-coded compositional types of Amiens frieze, south nave wall, second bay from the west

In fact, each bay has a unique sequence of compositional types and may be composed of anywhere from six to nine individual units between the pier sections that travel over the bundled shafts. On the outer transept walls, an overall pattern of alternating leaf forms and vine curls seems to have been desired, but this is achieved through a variety of block types instead of a uniform process.

The diversity of leaf forms within the first two sections of the frieze suggests experimentation and continuous refinement throughout the sculpting process. At Amiens, strict uniformity was not a priority in the execution of the frieze; in this instance, it appears that pattern and formal repetition were even abandoned in favor of a more organic aesthetic. It is possible that the usage of the stylistic outliers could also have been for economic reasons—perhaps the carved stone was too valuable not to be incorporated into the frieze and all attempts needed to be used in order to fill out the tremendous length of the nave and transepts, reducing the total amount of labor needed to complete it. But since the formal incongruities are present in every section of the frieze except the chevet, it is equally likely that uniformity and tightly controlled patterns were not the plan in the execution of the Amiens frieze; rather, sculptors sought to create an effect of organic plant life that had a sense of direction, movement, or growth that differed from earlier French examples, creating a new mode of architectural decoration for this specific iteration of Gothic design.

4.5 Dressing the Interior: Liturgical Considerations

Returning to the north nave wall and transept arms, it is curious that at various locations in the frieze, a vine curl is chopped off at the bottom abruptly, while other blocks feature a graceful vine curl that is visible from the bottom torus (fig. 4.48). Below the frieze, we notice a series of iron hooks inserted into the stone at regular intervals, much like the ones found at Reims (fig. 4.49). It appears that these hooks were intended for decorative textiles, as seen in the sixteenth century painting *The Mass of Saint Giles,* painted around 1500 (fig. 4.50). In the painting, the celebration of the mass takes place in the Gothic chevet of Saint Denis amidst numerous rich furnishings, including many examples of brightly colored textiles adorning the space around the altar. In addition, multicolored textiles hang just below the level of the triforium in the background (fig. 4.51). The hooks at Amiens and Reims likely displayed textiles in a similar manner, with various textiles that could be swapped out to reflect the progression of the liturgical calendar. A 1347 treasury inventory from Amiens indicates a number of silk panels intended for use during Holy Week, linen panels for Lent, and other types of fabric coverings [*cortinas diverse facture*] for the high altar.[199] Rotating the display of different textiles to mark the liturgical calendar was customary in Roman churches, documented as early as the fifth century and mentioned in *Liber pontificalis.*[200] In northern Europe, written documentation for church textiles includes Gregory of Tours, who describes their presence in the churches of Reims

---

[199] Raoul Rouvroy, *Ordinaire de l'église Notre-Dame, cathédrale d'Amiens*, ed. Georges Durand (Mémoires de la Société des antiquaires de Picardie, 1934), LV, XXVIII, XXIX.
[200] Laura Weigert, *Weaving Sacred Stories: French Choir Tapestries and the Performance of Clerical Identity* (Ithaca: Cornell University Press, 2004), 3.

during Clovis's baptism, as well as William Durandus, who mentions three distinct types of

church textiles: dressings for the church "in general," dressings for the choir, as well as coverings

and dressings for the altar. The textiles that adorn the church in general, according to Durandus,

include "curtains, canopies, and matching palls of silk and purple."[201] The portal sculpture of

Reims and Metz incorporates carved curtains into the lowest register of the portals, hanging by

fictive hooks spaced at regular intervals, a potential transmission of turning long, hanging

textiles into sculpted decoration.[202]

Examples of hanging textiles intended for the church walls do not survive at Amiens,

though it is notable that many surviving church textiles manufactured in the West before 1300,

the Bayeux tapestry and the Halberstadt tapestries, are composed in longitudinal formats suitable

for display in long stretches of space. At Amiens, the hooks are present below every section of

the frieze, including the nave, transepts, and the choir. The placement of these hooks would have

put the sculpted foliate frieze directly in conversation with these textiles, making it part of the

elaborate "dressing" of the church for liturgical functions. In this case, the frieze framed not only

the sill of the triforium, but the transition between the triforium and hanging textiles below, and

as such, the "chopped" vines would not have needed to continue past the liturgical dressings but

could be partially hidden behind them. This also indicates that the intention to dress the cathedral

with textiles was known to the sculptors and masons working on the cathedral in the 1230s.

In addition to the textile hooks, remnants of red ochre are visible on several sections of

the frieze and it has been presumed that the entire frieze was, at one time, painted in vibrant

---

[201] Weigert, *Weaving Sacred Stories,* 5. Honorius Augustodunensis and Rupert of Deutz also mention textiles as church dressings.

[202] Johannes Tripps, "Von der Kathedrale zu Reims bis zum Baptisterium von Florenz - Funde zum Behängen Mittelalterlicher Kirchenfassaden mit Textilien," in *Die Gebrauchte Kirche. Symposium und Vortragsreihe,* 2010, 83–89. The Reims curtains appear in the west portals; at Metz, the curtains are incorporated into the portal of Notre-Dame-la-Ronde.

colors. Faint traces of red ochre are visible on the surface of leaf forms and at least one fruit in

bay 7 of the north nave wall (fig. 4.52-4.54), but the surviving polychromy is most visible on

bays that abut the central crossing: in the northwest transept arm (fig. 4.55, 4.56), southwest

transept arm (fig. 4.57, 4.58), northeast transept arm (fig. 4.59, 4.60), and southeast transept arm

(fig. 4.61-63). In these examples, red ochre is visible in the negative space, or the "background,"

of the stones—whereas the visible polychromy in the north nave wall is located on the leaf forms

and fruits themselves. While it is possible that the entire frieze was painted—as many art

historians presume—the frieze might not have been painted in a uniform manner. The crossing

bays may have been renewed with new paint more frequently or possibly even emphasized in a

different manner. Additionally, the most visible surviving polychromy on the pier capitals also

appears atop the piers adjoining the crossing space, especially in the south transept. This raises

the possibility that the polychromy of the crossing and especially the south transept might

highlight the importance of this space and heighten processional entrances through the portal of

Saint Honoré, an entrance that once adjoined the cloister (now destroyed) and housing for

members of the chapter, serving as a clerical and important ceremonial threshold into the

cathedral's interior. While today's visitors typically enter the cathedral from the west,

progressing through the nave towards the east end, the canons who celebrated daily liturgical

services experienced the building entering through the south transept, processing directly into the

great crossing space. Heightening this alternative entrance through color and other decoration

suited the canons' experience and use of the architectural space and speaks to the priorities—

decorative, liturgical, and economic—of the cathedral's organization.[203]

---

[203] The central crossing of the church and its liturgical importance is explored in Murray, "Liturgical
Performance: Angels in the Architecture," *Notre-Dame of Amiens* (2021), 332-344. On liturgical celebration at
Notre-Dame of Amiens, see also Frédéric Billiet, *La vie musicale à Amiens au XVIe siècle* (Amiens: CNDP,

4.6 Observations on Leaf Forms in Medieval Art

A review of contemporary manuscript marginalia and illumination finds few foliate designs that replicate the particular combination of elements that appear at Amiens, though the simplified, tri-lobed leaf forms do share a certain aesthetic resonance with older foliate decoration found in Carolingian manuscript illuminations, such as those produced in the 8th-10th centuries in the scriptorium at the Abbey of Corbie. In some of the Corbie manuscripts, large, generalized leaf forms appear in conjunction with interlace ornament, some of which are tri-lobed (fig. 4.64, 4.65). Though it should also be noted that foliate forms in Carolingian manuscripts are remarkable in their diversity and generalized forms are sometimes supplanted by more naturalistic pointed leaf forms modeled closely on grape vines (fig. 4.66). Even so, there is not sufficient evidence that these forms were direct sources or even indirect inspiration for the frieze composition at Amiens. The likelihood of material transference from manuscript to sculptural decoration appears quite slim.

In the medium of carved stone, tri-lobed leaf forms are common in another early medieval art form: English crosses, where tri-lobed leaves frequently travel vertically up the sides of stone crosses with vine scrolls, abstractions that appear to stand in for grape leaves. Furthermore, the traditional definition of "stiff-leaf carving" in English architectural contexts includes tri-lobed leaf forms as a key characteristic of the corpus of English foliate carving; indeed, at Wells Cathedral, the foliate tympana of the west façade are rich with tri-lobed leaf forms similar to those found in the Amiens frieze and approximately contemporary to its

---

1984); Frédéric Billiet, "La maîtrise de la cathédrale d'Amiens du XIIIe au XVIIIe siècle," in *Amiens, la grâce d'une cathédrale* (Strasbourg: Editions La Nuée Bleue, 2012), 399–406.

manufacture (fig. 4.67). It is surprising similar examples of tri-lobed, vegetal carving are rather uncommon in France around or before 1220, making the appearance of these particular forms at Amiens all the more intriguing. While the formal similarity in the forms in the Amiens frieze and Wells façade is not sufficient to connect the two buildings, an English connection or source of inspiration cannot be ruled out at Amiens.

Villard de Honnecourt employs some trilobed forms in his drawings of leaf masks and a small section of a curving foliate frieze in folio 13 (fig. 1.7), but in terms of foliate carving from the period and especially foliate friezes from France, the simplified, graphic forms found in the Amiens frieze do not appear with great frequency in the region. More often, sculptors in France opted to use five-lobe or seven-lobe forms, sculpting elongated leaves with many appendages. In reviewing examples of foliate sculpture from the twelfth and early thirteenth century, the only comparable French leaf forms to the ones found in the frieze are found select aisle keystones at Amiens, which take on a number of foliate variations but include examples that incorporate tri-lobed leaf forms similar to those found in the frieze.

Furthermore, these leaf forms, unlike so many French examples from the first decades of the thirteenth century, do not correspond to any species of plant. The frieze foliage is a fantastical invention, though not any less complicated than examples of foliate sculpture that are more naturalistic. Instead of creating a monumental foliate element flush with naturalistic leaves, like the sculptors at Reims in the later capitals and field of lush, sculpted foliage behind the column figures of the west façade, the sculptors at Amiens chose to create an oversized element with abstract allusions to plant life. There were advantages to depicting foliage in this way; the viewer could fill in the identity of the plant, which could change according to liturgical need, whether

Eucharistic grape vine or celebratory garland fit for celebrating the Invention of Saint Firmin,

meaning shifting with the rotating liturgical textiles in the cathedral's interior.

The decision to carve foliage with no clear relationship to a naturalistic plant should be

considered in light of the plentiful examples of naturalistic grapevines and wild roses carved on

the western façade of the cathedral. Sculptors clearly had the ability to sculpt naturalistic vines

and leaves; why, then, are the leaf forms in the garland frieze rounded and out of proportion with

the accompanying fruits? At Amiens, the inability to definitively identify the plants in the frieze

as a grape vine also serves local hagiography and lore. A fantastical plant like those portrayed in

the frieze could stand in for a variety of iconographic meanings. With its tendrils and occasional

tangles of vines, the frieze could be read as a gesture towards Christ and the Eucharist. Its fleshy,

voluminous leaves, on the other hand, could also read as a garland of leafy foliage, appropriate

for the celebration of the Invention of Saint Firmin, with its miraculous blooms, leafy crowns,

and summer robes. While other carved foliage in the building is more direct in its botanical or

biblical references, it is worth considering whether the medieval sculptors emphasized its

potential for meaning by creating foliate forms that were plastic, organic, but malleable in the

eyes of viewers.

While some art historians of the nineteenth and twentieth centuries portrayed the frieze as

a vestige of a Romanesque past, even a cursory look at the forms of the frieze, its scholarship,

and its place within its architectural frame indicates that it is, in fact, an innovative and unique

aspect of the cathedral of Notre-Dame at Amiens and the vast material record of surviving foliate

sculpture in Gothic architecture. Comparison with other foliate friezes demonstrates that it was

well within the ability of medieval sculptors to produce a monumental, tightly controlled pattern

in architectural space. Following this discussion of foliate friezes and the questions that these sculptures raise in the context of medieval architecture, it is clear that the foliate frieze at Amiens diverges significantly from its earlier counterparts.

4.7 Conclusion

Sculptors knowingly departed from discernible patterns, introducing variations that surprise the eye and make the garland appear as if it is organically growing and bearing fruit, rather than reproducing itself mechanically. This moving depiction of growth not only captures the idea of a regenerative vegetative sculpture, but lays bare the artistic process. The frieze is playful and distinctive, a foil to the controlled geometry of earlier Gothic interiors at Noyon and Soissons. As such, the allowance and freedom granted in this frieze also presents an exploration of organic growth, rather than adherence to architectural trope or pattern. The purposeful liveliness of the frieze raises more questions about the manufacture of sculpted foliage in Gothic architecture and could serve as another means to glean information about medieval sculptural production at Amiens.

In the frieze at Amiens, however, the choice to incorporate a monumental foliate frieze with organic aesthetic qualities suggests the importance of natural metaphors in the construction of sacred space in northern France. The emphasis on representing plant life in permanent sculpture at Amiens speaks to the power and importance of this type of sculptural decoration, especially in its ability to shift and change with the liturgical calendar—a fitting visual for both Holy Week or the feast days of Saint Firmin. The fact that the frieze was composed to incorporate hanging textiles only amplifies this opalescent changeability and the lively, festival quality of the frieze.

While this analysis makes clear that sculptors chose to diverge from pattern and repetition, this need not be viewed as an inherently more advanced technique—indeed, no other cathedral or large church building project incorporated such a monumental frieze into its interior in France or England following the construction of the Gothic cathedral at Amiens. Gothic foliate friezes of the later thirteenth century are often dominated by crockets, or the alternating-leaf-crocket motifs seen in the chevet of Amiens, Sainte-Chapelle, the upper walls of Reims and elsewhere. If the frieze were to be considered in a framework of social competition and stylistic change, its singularity might be seen as an artistic failure; its tri-lobed forms "resisted" by nearby construction sites, not adopted, or elaborated upon, by other masons. But in terms of foliate sculpture, the Amiens frieze is not just an outlier, but a formidable accomplishment of unprecedented scale, unity, and organic beauty. Its singularity in France is an achievement that was unrealizable in other architectural contexts. In the first section of the frieze, a connection with the monumental foliate sculpture on the west façade of Wells remains a potential avenue of inquiry. In the third section of the frieze, the potential for exploring invention and transmission of repeating motifs in the latter half of the thirteenth century could be a fruitful direction for understanding the nuances of later trends in foliate sculpture, in which sculptors not only experimented with hyper-realistic forms, as seen on the chevet triforium capitals, but also opted for repetition and abstracted leaf forms in longitudinal friezes.

Pattern and repetition in architectural sculpture, liturgical textiles, and manuscript illumination remain largely unexplored topics in Western medieval art; scholarship in Byzantine and Islamic art can provide a way forward for prodding questions about the value of repetitive aesthetic choices, such as the floral diapering pattern that appears on the west portals of Notre-Dame of Noyon, Notre-Dame of Amiens, and Saint-Jean-des-Vignes in Soissons or the proliferation of the modular crocket, a topic that will be explored in the next chapter. Repetition in religious art and repetition in religious devotion are inherently linked; in addition to exploring the implications of sculptural production in creating repetitive forms, we might also consider the value of repetition in medieval culture more broadly, including its religious function in sacred space.

Finally, the frieze at Amiens provides an opportunity to think broadly about the implications of sculpting the natural world into the very stones of the cathedral, especially in a location where miraculous natural events were celebrated and even expected. Containing miraculous plant life within the cathedral carried with it certain advantages; like the figurative sculpture of the western portals, the sculpted foliage at Amiens was both palpable in a plastic, three-dimensional sense, and miraculous in its incomprehensibility. If Viollet-le-Duc and Durand, lifelong scholars of the cathedral who had deep, intimate knowledge of the building and its foliate sculpture, were not able to comprehend the frieze's forms, what effect did this frieze have on lay visitors and the clerical community who frequented the cathedral's interior? We might consider whether this frieze, brightly painted and further emphasized by hanging textiles below, became an example of miraculous foliage that could only blossom within confines of sacred space, representing miracles born in the natural world and the evergreen potential of resurrection. Matter with the potential for regeneration and change was a threat to the Church, as well as an opportunity. Environmental miracles could occur out in the fields, among the *petites plantes* that laypeople knew well and cultivated with great care. Sculpted foliage implicitly served to reinforce the power of the miraculous within the church, rather than out in the surrounding fields or vineyards. In this way, it was perhaps even advantageous to carve idealized, non-realistic foliage—leaf forms that appear nowhere in nature, not even in the keystones, capitals, or architectural crevices of neighboring churches. By staking a claim to

miraculous change, a cathedral that flowered forever would never face the corruption of death

and decay, as seen in the depictions of withered plants depicted in so many ways on Amiens'

west façade. Like the body of Saint Firmin found intact in his tomb and smelling of sweet,

miraculous odors, the eternally blooming foliage of Notre-Dame of Amiens reinforced the idea

of a sacred architecture created to reflect a constant state of grace.

5

The Crocket Capital and the Crocket Pier in England and France

The crocket, or *crochet*, is a hook-like appendage applied to a variety of surfaces in many architectural contexts across medieval Europe. While it is not exclusively foliate—it sometimes takes the form of a head or an animal—the crocket most often resembles a bud, leaf, or cluster of foliage and is often accompanied by other foliate elements. They can make a building appear to bristle, disrupt a sense of architectural geometry, and convey a sense of upward movement. Crockets enhance the play of light and shadow, and, like other forms of foliate sculpture, bridge architectural transitions.

According to the Oxford English Dictionary, the architectural meaning of the word has been in usage in English at least since the end of the fourteenth century, predating the use of Gothic/*Goti* by at least a hundred years.[204] This form appears in a dizzying array of variations in the architectural style we have come to refer to as Gothic: crockets can be found on pinnacles, gables, towers, friezes, capitals, and piers. It is, in essence, a modular unit that populates many kinds of architectural surfaces. This chapter will look at the use of crockets in Gothic architecture

---

[204] "Crochet," *Dictionnaire du Moyen Français,* Analyse et traitement informatique de la langue française and Université de Lorraine, <www.atilf.fr/dmf/>; and "Crochet," Ortolang, Centre national de ressources textuelles et lexicales, <www.cnrtl.fr>. According to the Oxford English Dictionary, the architectural meaning of the word 'crocket' can be traced to Middle English in the text of *Pierce Ploughman's Crede,* written in 1394. Earlier textual usages refer to a formal hairstyle, for which there are three fourteenth-century references. Later texts use the term in a variety of non-architectural ways, including in the description of musical notation beginning in the fifteenth century. In French, medieval texts from the 12th-14th centuries refer to a *crochet* as an iron tool, a hook or grapnel, or, in the late fourteenth century, the canine tooth of certain animals. 'Crochet,' as referring to the architectural form, is not common in French texts until the nineteenth century. It would appear, then, that these projecting elements were named in Middle English, but not in Middle French.

143

and, in reflecting on how they are used, consider an interpretation of a living architecture informed by its many uses and transformations.

5.1 Crockets as Foliate Sculpture

Understanding the use of crockets in Gothic architecture involves both a re-evaluation of the study of architectural sculpture and a return to the objects themselves, observing how transformations of this decorative scheme works *in situ*. When these sculpted elements are integrated into Christian architecture, they evoke biblical images such as the Living Cross, Joseph's Rod, Aaron's Rod, and the Tree of Jesse. These living stones mimic a stylized version of organic growth that has its roots in the classical past but has more flexible usage in Gothic buildings; two principal manifestations in England and France, the crocket "pier," or colonnette, as seen at Lincoln and the west façade of Wells, and the crocket capital, particularly at Noyon, Laon, Vézelay, and Paris, witnessed significant transformations in the twelfth and early thirteenth centuries. The significant variations of carved crockets provoke questions about the tension between naturalistic forms and geometry in medieval architecture, from the classic renditions at Noyon to the exuberant forms found in the nave of Wells Cathedral. The study of this deceptively simple phenomenon in specific buildings reveals the great flexibility of this modular system and its ability to transform the static materiality of stone into a living architecture organized around the principle of upward movement. These forms, while frequently commented upon (and frequently dismissed as derivative or formulaic), have not been the subject of a dedicated study since the publication of Viollet-le-Duc's *Dictionnaire raisonné*.

Villard de Honnecourt captured the general chalice shape of the Gothic crocket capital in his *cahier*, illustrating the towers of Laon and the interior elevation of Reims (fig. 5.1, 5.2). At

144

Laon, he captures not only the crocket capitals but the crockets that articulate the tower; at Reims, he draws crocket capitals crowning the compound piers of the interior in three dimensions. In both instances, Villard uses a generalized shape rather than copy the capitals themselves. In addition, the draftsman, or perhaps another subsequent owner of the *cahier,* appears to have begun to sketch a larger, more detailed drawing of a capital form flanked by two scrolls, but the sketch was never finished (fig. 5.3). Whether this was because the artist lost interest or was unable to complete the drawing due to lack of comprehension, we will never know. It is notable, however, that these elements are included in the schematized architectural drawings; crockets, in Villard's eyes, were understood as significant components of the building, even in a simplified representation.

5.2 Bud, Bloom, Decay: Crockets and Style

Writing about sculpture and architectural decoration in *Architecture and Society in Normandy 1120-1270*, Lindy Grant writes the following regarding the use of crockets:

> The passion of the classical probably had more impact on sculpture and on decorative details than on large architectural forms. But it lies behind the splendid variants on Corinthian capitals that graced most buildings until the dull, efficient crocket took over around 1180...[205]

This lament of the twelfth-century crocket takeover captures a certain boredom associated with this repetitive, modular form not uncommon in medieval architecture scholarship. The conventional understanding of crocket capitals, first suggested by Peter Kidson, has been that the rather pared-down design solved an engineer's dilemma, diverting valuable production time to the complexities of the structure rather than the details of capital sculpture. In contexts like the

---

[205] Lindy Grant, *Architecture and Society in Normandy 1120 to 1270* (New Haven: Yale University Press, 2005), 45.

Gothic chevet of Pontigny, however, crocket capitals introduced an element of movement and sculptural volume that is otherwise absent in the earlier phases of the building's construction, where masons employed flat cushion capitals and simplified moldings instead (fig. 5.4). Surely, master masons in charge of Gothic designs could have gone further to simplify capital forms if that was the intention, using cushion capitals or other simplified forms found in Cistercian architecture. Furthermore, crockets were not only employed in capital design, but attached to a variety of architectural elements such as cornices or jambs; in many cases, crockets heighten the aesthetic and decorative complexity of Gothic structures, rather than simplify or streamline them. The use and proliferation of crockets ushered in a new mode of architectural articulation that was quickly adopted and transformed many times over.

Crockets themselves have also been the vehicle through which historians of medieval art have illustrated theories of the origin of architecture and organic stylistic development. James Hall, writing in the late eighteenth century, looked to examples of domestic architecture built with simple, quotidian materials to explain the development of the pointed arch and foliate decoration in Gothic design.[206] Hall notes the abundant vegetal decoration of Gothic churches and posits that willow rods bound together to create pointed arches and pyramidical forms would, after some time, sprout, creating tufts of foliage along the form, forming interstitial growths (fig. 5.5). This is a possible explanation, he writes, why the Gothic churches he visited while traveling through France were covered with crockets and sculpted tufts of leaves; the sculpted decoration was meant to mimic these "primitive" building techniques, a thesis that reflects his knowledge of Vitruvian theories of the origins of architecture.

---

[206] James Hall, *Essay on the Origin and Principles of Gothic Architecture* (Edinburgh, 1797).

Viollet-le-Duc's *Dictionnaire* dedicates an entire entry to the subject of crockets and reprises the theme in his entries on capitals and sculpture. Over the course of these texts, the architect proposes theories on the role of the crocket in defining the break with antiquity, a theory of organic stylistic development, as well as hypotheses on the potential sources for these forms. Pinning the emergence of the crocket to the early twelfth-century construction at Vézelay, Viollet-le-Duc cites the interior capitals in the nave: "The interior capitals of the nave … already show us, in the place of the ancient volute, foliage turned back on itself, out of which come actual crockets."[207] Viollet-le-Duc uses this distinction of *feuillages retournés* to differentiate crockets from antique volutes and composite capitals, defining a break with, rather than continuation of, antiquity. Vitruvius described Ionic capital volutes as "hanging down at the right and left like curly ringlets and ornamented its front with cymatia and with festoons of fruit arranged in place of hair, while they brought the flutes down the shaft, falling like the folds in the robes worn by matrons."[208] The vegetal quality of these forms was, to Viollet-le-Duc, the important determinant in defining the break between volute and crocket.

Viollet-le-Duc also describes crocket cornices, expanding outside of the issue of capital sculpture to understand the crocket as a form onto itself. He dates the emergence of this phenomenon to the mid-twelfth century, claiming their presence signals another rupture, this time from Romanesque systems of architectural decoration. In short, Viollet-le-Duc's thesis frames the crocket as a harbinger of a new architectural style that breaks with the Romanesque and Classical past, an idea that serves his broader project of elevating and defending the Gothic style as innovative and rational. The analysis of the exemplars themselves, however, suggests

---

[207] "Les chapiteaux intérieurs de la nef…nous montrent aussi, à la place de la volute antique, des feuillages retornés sur eux-mêmes qui sort déjà de véritable crochets." Viollet-le-Duc, "Crochet," *Dictionnaire raisonné,* vol. 4, 400 (my translation).
[208] Vitruvius, *De architectura,* book 4, ch. 1:7, trans. Morgan, 104.

that this "break" was much more complicated; many early crocket capitals are difficult to differentiate from composite orders and Classical precedents, and sculptors appear to have chosen from a wide variety of crocket variation throughout the twelfth and thirteenth centuries.

In Viollet-le-Duc's theory of style, the transformation of the French crocket over time became a physical manifestation of the natural metaphor of organic stylistic development: the earliest crockets represent buds, then bloom into leafy clusters over the course of the thirteenth century, and eventually become overwrought and overcomplicated in their "decay." The architect argued the point so convincingly that this theory endured in the literature on Gothic architectural decoration, even as the organic theory of stylistic development lost traction more broadly in the field. In practice, however, sculptors opted for a number of crocket designs depending on the building site and use of the crocket; in the late thirteenth century, for example, when we might expect to find only blooming crockets looking like they are about to burst, according to the architect's theory, we instead find a number of instances in which sculptors elected to use the "bud" crocket form, sometimes in conjunction with their more vegetal counterparts. The development of crockets was not a linear narrative of organic growth, but a number of approaches to architectural articulation that varied from site to site. Perhaps because of the desire to see naturalistic foliage in Gothic architecture—and by extension a type of proto-Renaissance attention to illusionistic form—the repetitive types of foliate sculpture such as crockets have been characterized as derivative. If we re-frame repetition as a rhetorical device, however, rather than an economic or cost-saving measure, the proliferation of crockets in Gothic architecture provokes questions of intention and meaning in the design of sacred space.[209]

---

[209] Warren T. Woodfin, "Repetition and Replication: Sacred and Secular Patterned Textiles," in *Experiencing Byzantium: Papers from the 44th Spring Symposium of Byzantine Studies, Newcastle and Durham, April 2011*, ed. Claire Nesbitt and Mark Jackson (Surrey, England; Burlington, Vermont: Ashgate, 2013), 35–55. Woodfin

Like other forms of architectural decoration and foliate sculpture, the use of crockets

probes ideas of taste and propriety. Viollet-le-Duc was wary of buildings that employed too many

of these decorative modules: in establishing his typology of crocket variations, he is especially

disparaging of Ile-de-France construction between 1220-1230, a period during which, in his

estimation, crockets were overused:

> [In this period] the architect abuses the crocket: he puts it everywhere and uses it above
> all to articulate the straight lines which stand out against the sky, like the sides of
> steeples, the outer walls of towers, as can be seen at Notre-Dame of Paris or the bell
> tower of Senlis.[210]

While criticizing buildings that have a proliferation of crockets, he also expresses his distaste for

non-French variations. Here, he disparages the use of crockets in English architecture, a passage

that is illustrated by an image of one of the so-called Trondheim pillars, the vertical crocket piers

from Lincoln Cathedral:

> All the Anglo-Norman crockets of the mid-thirteenth century look alike, despite the
> efforts of sculptors who tried to put them into relief, a surprising model. They appear
> confused, and, at a distance, produce no effect because of the mass of the gouged "heads"
> and extreme thinness of the stems.[211]

Viollet-le-Duc sees France as the center of crocket production but did not approve of the increase

of its use in select French contexts of the thirteenth century; English sculptors, in his estimation,

were simply confused when they translated these forms into new architectural contexts. Their

addition to so many exterior architectural surfaces defied any sense of strict rationalism in

decoration, a reality that must have muddied the waters of Viollet-le-Duc's aim to rationalize

---

provides a model for studying repetitive art forms in sacred space and their potential for meaning in devotional use.

[210] "L'architect abuse du crochet: il en met partout, et s'en sert surtout pour denteler les lignes droites qui se détachent sur le ciel, comme les arêtiers des flèches, les piles extérieures des tours, ainsi qu'on peut le voir à Notre-Dame de Paris, au clocher de la cathédrale de Senlis." Viollet-le-Duc, "Crochet," 406 (my translation).
[211] "Tous les crochets anglo-normands du milieu du XIII siècle se ressemblent, malgré les efforts des sculpteurs pour leur donner du relief, un modèle surprenant, ils paraissent confus, et, à distance, ne produisent aucun effet, à cause de masse des têtes refoillées et de l'extrème maigreur des tiges." Viollet-le-Duc, "Crochet," 409 (my translation).

every element of the Gothic architectural system. Crockets served no structural purpose; in fact, as described in the first chapter of this dissertation, they were wont to break off in high winds or storms.[212]

The question remains as to why the crocket form was subject to so much resurgence and transformation in sacred architecture in England and France over the course of the twelfth and thirteenth centuries. Because of the simplicity of this form, we might pass over its importance; it is in their accumulation that crockets derive their aesthetic power. Even if that power is unsettling, as suggested by Viollet-le-Duc's analysis, confounding the architectural lines of the building or obscuring its geometry, the proliferation of crockets in sacred architecture is a notable phenomenon in medieval Christian architecture of this period.

5.3 The Crocket Capital and the Classical Orders

While it is not the purpose of this study to determine where the crocket first emerged, it is helpful to determine an approximate chronology to better understand the usage of the form in its medieval context and to identify its continuities and ruptures. Reflecting on the crocket's widespread use in the twelfth and thirteenth centuries and the flexibility with which builders adapted it into architectural systems provides a foundation for understanding its diversity and translation onto other architectural elements.

Many capitals that experiment with crockets and crocket-like forms have a visual relationship with composite orders that appear in late Antique and early Islamic contexts.[213]

---

[212] During Viollet-le-Duc's restoration at Amiens Cathedral, hundreds of crockets were re-made and replaced on the exterior of the building, and then replaced again as seasonal storms ripped through northern France and damaged those very restoration efforts.

[213] Richard Ettinghausen, *Islamic Art and Architecture, 650-1250* (New Haven [Conn.]: Yale University Press, c2001), 83-90; and Dale Kinney, "Roman Architectural Spolia," *Proceedings of the American Philosophical Society* 145, no. 2 (2001): 138–61, esp. 149-150; Rudolf Kautzsch, "Das korinthische Kapitell in Alexandria

While Early Christian basilicas often incorporated Roman spolia, capital carving remained a robust art form in Byzantine and Islamic architecture. The hypostyle hall at the mosque of Cordoba, for example, contains over eight hundred capitals atop spoliated columns; while the columns have been remarked upon at length, the capitals have come under considerably less scrutiny but are nonetheless an important body of architectural sculpture. Furthermore, the tenth-century palace Madinat al-Zahra outside of Cordoba houses a number of carved capitals that were signed by individual artists, signaling the importance of the craft and the form.[214] Tenth-century carved capitals from Cairo also point to experimentation, where composite orders gave way to a freedom of expression in foliate carving.[215]

In the Middle Ages, the capital became the site of much experimental sculpture, from the flat, lace-like foliate capitals found in many Romanesque churches and cloisters to historiated capitals displaying narrative scenes, such as those found at Vézelay and Autun. Despite their lack of figurative imagery, Gothic crocket capitals also experiment with traditional forms derived from composite orders and new compositions unique to particular buildings. The design systems of crocket capitals are, in fact, highly flexible and easily transformed through geometry and sculptural interpretation. The capitals that crown compound piers of varying shapes provide the transitional mechanism between the unique cross-section of the pier and the support of the wall

---

und Kairo," in *Kapitellstudien: Beiträge zu einer Geschichte des spätantiken Kapitells im Osten vom vierten bis ins siebente jahrhundert* (Berlin: W. de Gruyter & co, 1936), 24–39. While Dale Kinney highlights the use of spoliated columns in the first stage of construction at the mosque of Cordoba, she does not comment on the carved capitals. Medieval builders of the early Islamic empires often mixed re-used late Antique capitals with newly carved capitals, as extant examples in Cairo and Cordoba illustrate.

[214] Ettinghausen, *Islamic Art and Architecture, 650-1250*, 83. The signed capitals at Madinat al-Zahra appear in the chamber known as the Audience Hall. For a nuanced case study of the integration of Antique art into Islamic architecture, see Susana Calva Capilla, "The Reuse of Classical Antiquity in the Palace of Madinat Al-Zahra› and its Role in the Construction of Caliphal Legitimacy," *Muqarnas* 31 (2014): 1–33.

[215] Marianne Barrucand, "Remarks on the Iconography of the Medieval Capitals of Cairo: Form and Emplacement," in *The Iconography of Islamic Art*, ed. Bernard O'Kane, Studies in Honour of Robert Hillenbrand (Edinburgh University Press, 2005), 23–44

above, connecting to a respond or the springing of a vault and bringing this small element into

conversation with the larger, interconnected architectural system. In many cases, to study the

crocket is to examine its function within the integrated architectural frame.

In its many permutations, the crocket bears some resemblance to the classical volute; and

yet some of its more foliate variations tend towards acanthus leaf forms used in examples of

Corinthian capitals. The crocket capital is not an exemplar of any Classical order, however, and

the form's abundant variations are difficult to classify; as much as Viollet-le-Duc wanted to place

the crocket as a representation of a clean break with antiquity, its visual relationship to these

earlier forms is apparent in many examples.[216] Volutes and crockets belong to a shared lineage of

architectural decoration and coexisted in a dynamic network. For example, in Norman building

projects of the second half of the eleventh century, masons were employing a volute-like unit in

capital carving, as at St. Etienne at Caen, where the volutes emphasize the diagonals of the

square abacus, supported by a second row of flat, leaf-like elements akin to a composite capital

(fig. 5.6).[217] At Ely Cathedral, Norman capitals include these scrolls combined with foliage

carved in flat relief (fig. 5.7, 5.8). Notably, this foliate carving and scrolls were not integrated

into the nave capitals, where flat capitals bridge the transition from pier to supporting arch, a

change that might have been related to the gap in leadership during the construction at Ely, then

an abbey church.[218]

---

[216] Most Gothic capitals forego figurative sculpture, though exceptions can be seen both at Wells Cathedral and Notre-Dame of Reims. Other instances of older capital designs worked into the fabric of the building can also be seen at Saint-Denis, Noyon, Bourges, and Notre-Dame of Paris.

[217] Maylis Baylé, *La Trinité de Caen : sa place dans l'histoire de l'architecture et du décor romans* (Genève: Droz, 1979). Similar forms can be seen in Église de la Trinité, commissioned by William the Conquerer's wife Matilda, though the decorative sculpture in the capitals and moldings is notable for its diversity and variation, containing many patterns, animals, hybrids, and figures.

[218] J. Philip Mcaleer, "Some Observations about the Romanesque Choir of Ely Cathedral," *Journal of the Society of Architectural Historians* 53, no. 1 (1994): 80 94; Sarah Ferguson, "The Romanesque Cathedral of Ely: An Archeological Evaluation of Its Construction" (Doctoral Dissertation, Columbia University, 1986); and George Zarnecki, *The Early Sculpture of Ely Cathedral.* (London, 1958), pp. 6-12. Simeon, a monk of St.

5.4 The Crocket Capital in France: Physical Evidence

Given more flexibility in defining the formal manifestations of the crocket, we might

consider a longer history of the form and a broader definition of its prototypes to understand its

emergence in Gothic architectural systems. In certain examples the carver appears to experiment

with prototypes of the crocket form in the early and middle decades of the twelfth century: At

Saint-Germer-de-Fly, begun in the 1130s, a variety of forms in the ambulatory pier capitals,

some of which bear similarity to Norman examples found at Caen (fig. 5.9). Other capitals,

however, have longer, leaf-like forms that reach out towards the corners of the abacus, the

pointed tip of the leaf pointing outwards, foregoing the scroll element. A similar decoration

scheme can be found in several French churches of the mid-twelfth century.

In conjunction with Romanesque forms in the capitals on the outer periphery of its

ambulatory, the chevet at Notre-Dame of Noyon, begun around 1145, employs flat, leaf-like

capitals in the hemicycle.[219] These capitals, placed atop round piers, reach up to a square abacus

(fig. 5.10). In this form, the large leaf unites the diagonals of the abacus, while a shorter unit

---

Ouen at Rouen and brother of the bishop of Winchester, was appointed the abbot of Ely in 1081, where construction on the new church began during his tenure. By Simeon's death in 1093, the transepts and choir may have been well underway; a new abbot, Richard, was not appointed until 1100. Richard campaigned for Ely to become a new bishopric, which was not realized until a year after his death, in 1108. The transept arms at Ely were largely rebuilt after the collapse of the Romanesque tower in 1322, which accounts for many dissimilarities in the building's fabric. Late eleventh-century remnants were, however, incorporated into the rebuilt sections, along with fourteenth-century additions.

[219] Arnaud Timbert, and Philippe Montier. *La cathédrale Notre-Dame de Noyon et son quartier.* Noyon: Cap régions, 2012. The capitals at Noyon have been employed in sorting out the complicated issue of construction chronology (for an example, see Mark Pessin, "The Twelfth-Century Abbey Church of Saint-Germer-de-Fly," Gesta, XVII, 1978, 71), even though capitals such as these, as in many late medieval buildings, are likely to have been sculpted avant-la-pose and put in place long after they were carved. The placement of capitals and the changes in capital carving should be observed separately from building chronology to avoid erroneous conflation. Inferring additional information about the capital from other chronology evidence can shed light on the construction process, but architectural historians are now wary of using capitals as evidence for parsing out construction sequences.

reaches up the axial planes of the capital. In the nave, this combination is used again, the terminus of the leaf units curling under to form a recognizable crocket, which is neither a scroll nor a broad "leaf" with a pointed tip (fig. 5.11). The proto-crocket is further elaborated upon in some examples, where a third row of crockets is introduced. On the north side of the nave, certain capitals introduce an independent leaf element below three rows of crockets sprouting out of the top half of the capital, demarcated by a gentle molding that recalls later examples from the nave of Soissons (fig. 5.12). Often these leaf elements are idealized or generalized forms or incorporate a pearling detail on its central stem, but other examples incorporate leaves from recognizable species like oak or grape vine.

In the latter half of the twelfth century, early versions of crocket capitals are often combined with examples that feature flat, leaf-like projections. Both types of forms are found in the chevet at Vézelay, built approximately between 1165-1180, the first architectural design of its kind in Burgundy (fig. 5.13).[220] The four-level south transept at the cathedral of Soissons constructed around 1170-1180 integrates many examples of French early crocket capital design in the hemicycle.[221] Above the hemicycle's slender columns sit capitals with acanthus leaves as

---

[220] Francis Salet, *La Madeleine de Vézelay* (Melun, France: Librairie d'Argences, 1948), 28; Dieter Kimpel and Robert Suckale, *Die gotische Architektur in Frankreich 1130-1270* (München: Hirmer, 1995), 145; Arnaud Timbert, *Vézelay : le chevet de la Madeleine et le premier gothique bourguignon* (Rennes: Presses universitaires de Rennes, 2009). The chevet at Vézelay represents an important chapter in Gothic design. Following evidence presented by Dieter Kimpel, Robert Suckale, and Arnaud Timbert, the beginning date of construction was determined to be around 1165, twenty years before the chronology proposed by Francis Salet's 1948 monograph. The visual relationship between these early crocket capitals to examples in other early Gothic buildings, such as those found in the nave of Noyon, lends further support to this theory.
[221] Dany Sandron, *La cathédrale de Soissons : architecture du pouvoir* (Paris: Picard, 1998), 27, 75-76. A fragment of an engaged capital from the Romanesque church with carved acanthus designs was discovered at Soissons in 1924. Carl Barnes hypothesized that the capital once belonged to the previous iteration of the cathedral's west façade. The integration of acanthus motifs in the south transept hemicycle may have connected to other sculpted elements in the south transept, serving either as an inspiration of motifs or a vehicle for visual continuity. Sandron observed this duality in the design of the south transept capitals; while those in the upper chapel likely belong to later restoration efforts, the hemicycle capitals are important for the study of architectural carving in the Gothic system.

well as more restrained forms (figs 5.14-15). In the nave of Notre-Dame of Laon, where construction proceeded from east to west and concluded before 1200, the capitals crowning the round piers of the nave use both of these forms (figs. 5.16-17).[222]

Notably, crocket capitals in early Gothic structures are frequently placed atop piers in the hemicycle, contrasting with Romanesque or Corinthian acanthus leaves in the outer ambulatory capitals. Such an arrangement can be seen at Noyon as well as Saint-Denis, where capital sculpture has been the subject of many detailed studies.[223] In the chevet at Saint-Denis, bi-level crocket capitals appear on the round piers of the hemicycle, likely replacements dating to the rebuilding of the upper choir, while acanthus-type foliate forms reminiscent of late Antique and Romanesque carving decorate the capitals crowning the monolithic columns of the ambulatory (fig. 5.18). At Saint-Denis as well as Noyon, the chevet introduces architectural forms associated with a burgeoning new system of architectural design.[224] This juxtaposition presents a dialectic

---

[222] Éric Fernie, "La fonction liturgique des piliers cantonnés dans la nef de la cathédrale de Laon," *Bulletin Monumental* 145, no. 3 (1987): 257–66. There are also four compound piers alternating with round piers in the eastern bays of the nave at Laon. These piers have capitals with smaller, integrated crocket forms on the colonnettes and the central round pier. While it is thought that this alternating scheme was simply abandoned in favor of the round piers as construction proceeded on the nave, Fernie posited that the decision to highlight the eastern bays architecturally was liturgical in motivation.

[223] William Clark, "'The Recollection of the Past Is the Promise of the Future.' Continuity and Contextuality: Saint-Denis, Merovingians, Capetians, and Paris," in *Artistic Integration in Gothic Buildings*, ed. Virginia Chieffo Raguin, Kathryn Brush, and Peter Draper (University of Toronto Press, 1995), 92–113; May Vieillard-Troiekouroff, "Les chapiteaux de marbre du haut Moyen-Âge à Saint-Denis," *Gesta* 15, no. 1/2 (1976): 105–12. The crocket capitals in the hemicycle at Saint-Denis are likely replacements dating to the rebuilding of the upper choir. The juxtaposition between the Merovingian carved capitals in the crypt and the reimagined forms found in the outer aisle of the chevet have been the subject of several studies of considerable depth, though the pattern of juxtaposing "older" forms in the aisle with "newer" forms around the hemicycle is clearly a pattern in twelfth-century church building in France, as seen at Noyon, Saint-Denis, and Notre-Dame of Paris. The intense study of the capitals at Saint-Denis can provide a model for investigating the apparent purposeful variation in capital carving in Gothic space.

[224] Alain Erlande-Brandenburg, "Le choeur dans les cathédrales gothiques" and "La conception spatiale des chevets gothiques : points de vue liturgiques," in *La place du choeur: architecture et liturgie du Moyen Âge aux Temps modernes: actes du colloque de l'EPHE, Institut national d'histoire de l'art, les 10 et 11 décembre 2007*, ed. Sabine Frommel, Laurent Lecomte, and Raphaël Tassin, Itinéraires percorsi 1 (Paris : Rome: Picard ; Campisano, 2012): 57-66, 67-78; Stefaan van Liefferinge, "The Hemicycle of Notre-Dame of Paris: Gothic Design and Geometrical Knowledge in the Twelfth Century," *Journal of the Society of Architectural Historians* 69, no. 4 (2010): 490–507. While the form of the chevet was employed in Romanesque churches associated with pilgrimage, this form was crucial to the development of Gothic building systems, the building

between old and new in the heart of the building's interior as well as a thread of continuity with the architectural past.

In light of these examples, the crocket capitals at Notre-Dame of Paris are notable for their diversity and unique in their elaboration.[225] In the chevet largely completed by 1177, the sculptural field features several different crocket forms, combining crocket forms, scrolls, and broad, flat-leaf forms in exuberant combinations that appear to multiply and spill over (fig. 5.19).[226] These capitals highlight negative space; the crockets curling over are deeply undercut areas. While other early Gothic capitals emphasize the diagonals of the structure and geometric order, the stout Parisian capitals have been composed in such a way as to produce an effect of multiplicity. While the individual parts of a capital at Noyon, Saint-Denis, or Vézelay are easily understood, sculptors at Notre-Dame of Paris playfully overlapped forms in a variety of patterns in ways that disguise or even muddle its individual parts. Most of the capitals are carved in two

---

of which was informed by liturgical need, geometrical knowledge, and a desire for a new architecture.

[225] William W. Clark, "The Early Capitals of Notre-Dame de Paris," in *Tribute to Lotte Brand Philip: Art Historian and Detective*, ed. William W. Clark and Lotte Brand Philip (New York: Abaris Books, 1985), 34–42; Denise Jalabert, "La première flore gothique aux chapiteaux de Notre-Dame de Paris," *Gazette des Beaux Arts*, May 1931, 283–304. Jalabert searched for naturalistic plants in the carved foliage of the capitals at Notre-Dame of Paris, remarking on the monumental quality of the carved foliage. Clark highlighted the Romanesque figures and acanthus forms in the capitals of the outer ambulatory, comparing their forms to similar carvings at Saint-Germain-des-Prés and, ultimately, suggesting that construction on the cathedral began in the 1150s. The group of diverse crocket capitals, however, have not been examined with the same intensity.

[226] The construction of Notre-Dame of Paris was a dynamic field of change that remained fluid and open to revision throughout the Middle Ages. The bibliography related to the issue of building chronology at Notre-Dame of Paris is vast, but see particularly Dany Sandron and Andrew Tallon, *Notre Dame Cathedral: Nine Centuries of History*, trans. Lindsay Cook (University Park, Pennsylvania: Penn State University Press, 2020); Clark, "Early Capitals at Notre-Dame," 41-42; Caroline Bruzelius, "The Construction of Notre-Dame in Paris," *The Art Bulletin* 69, no. 4 (1987): 540–69; Mailan S. Doquang, "The Lateral Chapels of Notre-Dame in Context," *Gesta* 50, no. 2 (2011): 137–61. The date of the start of construction of Notre-Dame is still somewhat contested; the oft-cited date of 1163 is based on a tradition that Pope Alexander III visited Paris in that year and laid the first stone; this account appears to originate from an fourteenth-century chronicle by Jean de Saint Victor, over a century after the event, and is not corroborated by surviving documentary evidence related to Pope Alexander III's travels during his tenure. A written account attributed to Robert de Thorigny, abbot of Mont-Saint-Michel, reported that the chevet was largely finished in 1177. Work on the nave continued into the thirteenth century, with the nave elevation mostly finished by 1210. Construction on the lateral chapels in the nave began in the 1220s and further changes to the cathedral continued for decades.

levels—though in some designs, this boundary is de-emphasized—and draw attention to the diagonals of the square abacus, sometimes with crockets that beget smaller crockets just above them.

At Notre-Dame of Paris, there are also outliers: in the inner ambulatory, one capital with linear acanthus forms is a particularly effective foil to the deeply undercut capitals in the chevet (fig. 5.20). This group of capitals has formal similarities to carvings at Saint-Germain-des-Prés and appears to have been executed before 1160.[227] This pattern of older forms in capitals in the outer ambulatory can be seen in a number of Gothic buildings and, while their presence may be interpreted as evidence that construction began on the outer wall, we should also consider the architectural dialectic that the two modes of carving present in the dynamic, curving space of the chevet. Notre-Dame of Paris, like Noyon and Saint-Denis, employs the crocket capitals—and not hybrid forms, masks, or animals—in the hemicycle, the architectural frame for the high offices of the Church. Figurative sculpture, Romanesque foliate carving, and classicizing acanthus leaves were, in these cases, was saved for the margins of the interior space, relegated to the periphery.

A similar change can be observed in the gallery capitals (fig. 5.21, 5.22). In the chevet, capital carving follows linear acanthus models more commonly associated with Romanesque architecture. While these capitals emphasize the diagonals and allow the acanthus leaf to fold over and project into space, they are a contrast to the crocket capitals in the gallery in the nave.[228] In the crocket capital, the forms change to emphasize negative space; the chalice shape is notable for its extreme curvature and projecting crockets, which pull the forms outward. The effect is that

---

[227] Clark, "The Early Capitals of Notre-Dame de Paris," 34–36.

[228] Many of the gallery crocket capitals in the nave appear to have been replaced during the nineteenth-century restorations of the cathedral, adding to the complexity of what type of foliate carving was deemed correct for this part of the building. While the restored capitals follow standard formulae, no two are identical. In the chevet, at least one gallery capital appears to be smaller manifestations of a capital in the hemicycle, but most employ acanthus leaf forms that hug the conical shape.

the capital appears liberated from its conical shape, the abacus balancing on ever more

precarious footing: the springing of the arch balances not on a conical capital, but on a few sprigs

of fantastical foliage.

The pier capitals in the nave of Notre-Dame of Paris, particularly those in the easternmost

bays, present many variations of the crocket-foliage model of carving (figs. 5.23-5.26). Like the

pier capitals in the chevet, each example is a unique combination of foliate forms and crockets.

In many of the nave capitals, three- and five-point leaf forms not only alternate with the crockets,

but overtake the entire decorative scheme, producing a lace-like texture with acanthus leaves that

nearly obscures the crocket form altogether. In the first and second westernmost bays that

employ engaged shafts, the sculptors' response to the changing form shows a shift in the

treatment of foliage. In the second bay, the capital is interrupted by the engaged shaft; on the

north side, the engaged shaft employs its own decorative schema; on the south, the shaft's design

has been integrated into the entire capital (fig. 5.25). In the first bay, however, which employs

four engaged shafts in a *pilier cantonné,* the crocket form has nearly been abandoned in favor of

all-over foliage, comparable to similar examples at Reims. Small crockets make an appearance

on the south pier at the springing of the arch, but the pier makes use of diagonal foliate forms

that appear tousled by the wind, recalling a type of classical windblown acanthus capitals found

in Damascus and Thessaloniki (fig. 5.26).[229]

At the cathedral of Saint-Gervais-Saint-Protais in Soissons, the delicacy of the carving of

the capitals in the south transept stands in sharp contrast to the robust, stately capitals that crown

the substantial round piers in the chevet and nave (figs. 5.27-30). Like Notre-Dame of Paris, the

chevet contains a wide variety of examples, from rows of diagonally set leaves—not unlike the

---

[229] Rudolf Kautzsch, "Kapitelle mit schräggestelltem und umgewehtem Akanthus," in *Kapitellstudien,* 140–52.

westernmost pier capital at Notre-Dame of Paris—to bilevel rows of alternating crockets. In the nave, bilevel crocket forms are standardized to coordinated with octagonal abacus, though sculptors continued to experiment with the placement of the middle band. In the westernmost pier in the south arcade of the nave, one capital employs flatly carved foliage alternating with crockets (fig. 5.30), a departure from the rest of the restrained crocket capitals atop piers and sculptural gesture that, in its use of carved foliage, points to the nave capitals at Notre-Dame of Paris.

At Notre-Dame of Chartres, sculptors used crocket forms and single leaf motifs to unify the capitals of the *piliers cantonnés*, integrating the decoration of capitals atop the engaged shaft with the whole (figs. 5.31-33).[230] The crockets are accompanied by naturalistic foliage, including identifiable oak leaves and grape vine leaves. These capitals standardize the crocket to an almost classicizing degree, maintaining strict order and symmetry. Some piers and shafts at Chartres make use of an octagonal abacus, which corresponds to crockets emphasizing each corner. For the most part, capitals at Chartres alternate crockets with leaf forms in a bilevel design, with the engaged shaft capital beginning just above though some capitals in the chevet experiment with rows of crockets and bunches of foliage clustered together.

In the first decades of the thirteenth century, sculptors experimented with several solutions when adorning the capitals atop *piliers cantonnés*. In the earlier capitals in the eastern bays of the nave at Notre-Dame of Reims, crockets co-mingle with flat leaf motifs, effusive

---

[230] In an architectural ensemble so rich with extant sculpture and stained glass, it is not surprising that the restrained crocket capitals of Chartres have rarely caught the attention of art historians. The decision to utilize these restrained forms, however, in conjunction with portals and stained glass rich in imagery, texture, and color, is notable and might reflect the decorative priorities of this particular building. An investigation of foliate carving at Chartres could shed light on decorative priorities in Gothic design, particularly the degree to which these decisions were individualized in local *chantiers*.

foliate carving, and figurative subjects (fig. 5.34-36).[231] The figurative subjects appear

exclusively on the engaged shafts in an additional sculptural field included below the shaft

capital. As at Chartres, the abaci of the engaged shafts connecting to the springing of the arches

are octagonal, encouraging a proliferation of crockets at each corner. One capital at the eastern

end of the north aisle integrates earlier single-leaf forms, connecting to twelfth-century examples

at Noyon and Laon (fig. 5.30). In several later examples at the west end of the nave, the crockets

are—as at Notre-Dame of Paris—overtaken by the foliage and sometimes disappear altogether

(fig. 5.35). Some of these capitals are split into four levels of detailed, deeply undercut foliage, a

hyper-realistic mode of foliate carving also employed in concert with the portal sculpture of the

west façade (fig. 5.36).

At Notre-Dame of Amiens, crockets mingle with leaf forms in the *piliers cantonnés* (fig.

5.37). These capitals bear similarities to those found in the nave at Chartres, though the sculptors

at Amiens integrate the motifs fully, overlapping leaf forms with crocket forms to a degree that

makes them difficult to differentiate, especially from a distance. The foliate motifs travel across

each capital, fully integrating the engaged shafts into the ensemble. But the sculptors appear to

have experimented a great deal to arrive at this scheme: In the earliest pier capital in the

easternmost north bay of the nave, an unfinished, chiseled ridge reveals the adherence to a

bilevel design, much like examples from Chartres and Soissons (fig. 5.38).[232]

---

[231] Barbara Abou-El-Haj, "Building and Decoration at Reims and Amiens," in *Europäische Skulptur Im 12./13. Jahrhundert*, ed. H. Beck and K. Hengevoss-Dürkop, vol. I, 2 vols. vols. (Frankfurt am Main, 1994), 763–76; and "The Lordship and Commune Project," International Center of Medieval Art, accessed July 15, 2020, Abou-El-Haj pointed to the archbishop's control over funds at Reims in order to explain the "lavish" decoration of the cathedral. While the history of the political economies in both locations is of great importance, her characterization of the decorative carving at Amiens as "simplistic" and "serialized" is inconsistent with the actual extant carving at Amiens, which is notably diverse in its sculpted decoration. The 1994 article became the basis for a larger comparative study that was unfortunately left unfinished at the time of Abou-El-Haj's untimely death in 2015, which might have shed light on these issues.

[232] Murray, *Notre-Dame, Cathedral of Amiens* (1996), 44-58. In the case of Amiens, the experimental capital

Smaller capitals in the dado and nave triforium adopt simple forms on a reduced scale (figs. 5.39-41). The thirteenth-century dado at Amiens is in poor condition, having been retrofitted with various additions to the cathedral that have resulted in snapped-off crockets and damaged sculptural details (fig. 5.41). When the lateral chapels were added to the sides of the nave, the original dado was incorporated into the outer wall, now hidden behind wood paneling. In the thirteenth century, the dado would have been an eye-level arcade of many variations of sculpted foliage, like examples at Lincoln and in the quire aisles at Wells. While the sculpting of pier capitals provides important examples of large-scale foliate sculpture, the small sculpture of the dado capitals would have been more accessible to cathedral visitors; the visibility of this foliage could activate the many other foliate details in the building, such as the great foliate band that encircles the cathedral's interior. In the nave triforium, capitals feature several variations of the alternating crocket forms. In comparison with examples from the gallery of Notre-Dame of Paris, the sculpted foliage is less delicate in execution, though some capitals bear traces of polychromy (fig. 5.42). What is perhaps most curious at Amiens is the carved foliage in the St. Christopher portal, an entrance dating to the 1220s that might have been constructed to service the south aisle while the western portals were being completed (figs. 5.43, 5.44). The foliage in these capitals is far more delicate than the dado or the triforium; the foliate capitals to the left of the portal are entirely covered with foliage with crocket forms (most of which have broken off), while the capitals on the right side of the portal have a more defined crocket structure overlayed by naturalistic leaves carved in shallow relief. These capitals also incorporate a carved abacus and sculpted torus; the attention to detail in these examples suggests that this type of delicate foliage might have been intended for the rest of the cathedral. This portal is crowned by a tri-

designs can be found in the bays closest to the start of construction.

lobed tympanum of the floral diaper relief that articulates the bottom registers, trumeau posts, and lintels of the west portal, further connecting the temporary portal with the earliest phases of construction and the cathedral's overarching program of foliate motifs (fig. 5.45).

In France, crocket capitals were subject to wide variation and experimentation. Variations often co-existed within several *chantiers,* suggesting a preference for variety in capital carving as well as ongoing iteration within local workshops. As these examples show, the juxtaposition of new and old sculptural forms appears in distinct patterns. In many French contexts, crocket forms were chosen for capital carving in updated Gothic designs in the hemicycle, a space that usually coincided with the placement of the high altar. That this type of abstract foliate carving was chosen over modes, such as acanthus leaves or narrative scenes, is notable. Rather than a takeover of dull repetition, crocket forms provided a new direction for architectural sculpture in the dynamic, quickly changing architectural system.

5.5 The Crocket in England: Modular Transformations and Stiff Leaf Foliage

When surveying capital carving in Gothic building projects in England, art historians have often used the term "stiff leaf" to classify a wide array of foliate forms. There is reason to consider this type of foliate sculpture within the context of the experimentation occurring on the Continent not only because of its contemporality, but also because of the ways in which masons in England transformed the modular system of crockets onto other architectural elements in the Gothic system.

Stiff-leaf foliage, considered by Samuel Gardner to be "the most perfect expression of our exclusively English Gothic foliage sculpture," has long been considered a unique category of foliate sculpture by scholars of English medieval architecture.[233] First coined by Frederick

Moore Simpson in 1909, the term 'stiff leaf' entered the discourse around medieval sculpted

foliage more fully with the publication of Gardner's short volume on English Gothic foliage, in

which he devoted a chapter to the topic.[234] To date, the most extensive study of stiff-leaf foliage

is the 1952 dissertation written by Pamela Tudor-Craig, who observed that the term itself had

rarely been explained in an analytic manner and proceeded to lay out several theories as to its

stylistic development and seemingly infinite transformations.[235] In general terms, "stiff leaf" has

been defined as the use of trefoil leaf forms in conjunction with an emphasis on stiffly carved

stalks. This particular mode of carving, along with hyper-naturalistic examples, such as those at

Southwell, are often considered the two most important "moments" in the history of English

Gothic foliate carving. Opening up the study of foliate sculpture in England reveals, as in France,

that sculptors and masons were embedded within a dynamic web of decorative possibilities and

chose the crocket, in particular, to develop many modes of architectural articulation.

After establishing "stiff leaf" as a national English phenomenon, Tudor-Craig's

dissertation explored the impact of French capital design in English buildings, referring to the

"French classic acanthus" model of capital used in the Trinity Chapel of Canterbury Cathedral,

citing examples at St-Denis, Le Mans, Tours, and St-Remi in Reims—buildings that employ, in

addition to acanthus leaves, a variety of foliate forms in capital carving. Following the murder of

Thomas Becket in 1170 and a disastrous fire in 1174, master mason William of Sens was charged

with rebuilding the quire at Canterbury Cathedral. But, as Gervase of Canterbury recounts, the

---

[233] Samuel Gardner, *English Gothic Foliage Sculpture* (Cambridge, Eng.: The University Press, 1927), 2.
[234] Pamela Tudor-Craig, "Stiff-Leaf," Grove Art Online, accessed July 17, 2020; F. M. (Frederick Moore) Simpson, *A History of Architectural Development*, vol. 2 (London, New York, etc.: Longmans, Green, and Co, 1909), 128; Gardner, "Early English Stiff Leaf," in *English Gothic Foliage Sculpture*, 22-32.
[235] Pamela Tudor-Craig [Pamela Wynne-Reeves], "English Stiff Leaf Sculpture" (Ph.D., University of London (Courtauld Institute of Art), 1952).

first William suffered an unfortunate fall in 1177 that left the architect badly injured and, though

he tried to direct work while bedridden, he was succeeded by William the Englishman. The first

William died in 1180, while the second William carried on with construction, building eastward

to complete the vaults and the Trinity Chapel.

At Canterbury, the differences in capital carving between the work associated with

William of Sens and William the Englishman are clear: in the choir, capitals are carved in low

relief, the acanthus leaves gently curling outwards at the corners of the abacus and the axes of the

pier (fig. 5.46). In the Trinity Chapel, the capitals follow this formula but exaggerate the negative

space, pulling the ends of the foliate forms, now stems with multiple individual leaves, upward

and curling in on themselves (fig. 5.47). In the space between these two modes of carving are the

capitals in the crossing atop compound piers with Purbeck marble shafts; these capitals

incorporate recognizable crocket forms, including the now-familiar crocket pattern with a single

alternating leaf, which can be seen in the capitals on the eastern crossing piers (fig. 5.48). In a

portion of the building where the old work is reconciled with the new, these crocket capitals

bridge the space between the work of the two masons, offering a thesis-antithesis-synthesis that

results in a bulbous, foliate capital whose exaggerated volume and negative space is yet another

manifestation of the crocket phenomenon.[236] Gervase of Canterbury hints at these observations,

referring to the "fine carving" in the quire and emphasizing the importance of the four "principal

---

[236] Peter Draper, "Patrons, Masons and Saints: Canterbury Cathedral," in *The Formation of English Gothic : Architecture and Identity* (London: Published for The Paul Mellon Centre for studies in British Art by Yale University Press, c2006), 13–34; Stephen Murray, "Gervase of Canterbury: *Cronicus* and Logistics Man," in *Plotting Gothic* (Chicago: University of Chicago Press, 2014), 47–71, esp. 53-63; Gervase of Canterbury, *Of the Burning and Repair of the Church of Canterbury : In the Year 1174, from the Latin of Gervase, a Monk of the Priory of Christ Church, Canterbury* (Cambridge: Cambridge University Press, 1932). Gervase of Canterbury (1141-c1210) provides an invaluable medieval account of this phase of construction, focusing in detail on the comparison between the old work and the new. This dialectical approach informed many studies devoted to the cathedral's architecture.

pillars" surrounded by Purbeck marble, evidence that he was attuned to variations in foliate carving and invested in the architecture of the piers at Canterbury.[237]

Some of the earliest surviving crocket designs in England, however, are found in the early phases of construction at Wells Cathedral, where sculptors experimented with incorporating a variety of bulbous, crocket-like forms in the three-bay choir constructed between 1174-1184 (fig. 5.49, 5.50).[238] These capitals, perhaps because of their irregular execution, have rarely been commented upon despite being from one of the earliest construction sites of the Gothic building enterprise in England.[239] The easternmost piers present a single register of curling leaf forms— some trefoiled, some more complex—with some corners emphasized by a second leaf form appearing above the main curling form, as at Notre-Dame of Paris. Atop the piers to the west, these projections split into two registers, a more standardized approach that belies certain irregularities in the levels and execution (fig. 5.51, 5.52).

---

[237] Gervase of Canterbury, *Of the Burning and Repair of the Church of Canterbury,* 17-18. Abbot Suger also has a good deal to say about the monolithic columns that make up the hemicycle of Saint-Denis, including a miracle story of their origin and a discussion of their symbolism.

[238] Jerry Sampson, *Wells Cathedral West Front: Construction, Sculpture, and Conservation* (Phoenix Mill, Thrupp, Stroud, Gloucestershire: Sutton Pub, 1998), 13, 16-17; Warwick Rodwell, *Wells Cathedral: Excavations and Structural Studies, 1978-93* (London: English Heritage, 2001); G. A. A. Wright and W. A. Wheeler, *Masons' Marks on Wells Cathedral Church,* 2nd Revised edition (Wells: Friends of Wells Cathedral, 1971); Carolyn Marino Malone, *Façade as Spectacle: Ritual and Ideology at Wells Cathedral* (Boston: Brill, 2004). Bishop Reginald began his post at Wells in 1174 and began planning the rebuilding of the church shortly thereafter, it is thought that the first two building phases consisting of the Lady Chapel, choir, and eastern aisles of the transepts was completed around 1184, given the change from Doulting stone to Chilcote at this juncture. Doulting stone came from a quarry owned by Glastonbury Abbey, which suffered a fire on May 25, 1184. Chilcote stone was quarried near the cathedral. Construction continued westward, the western bays of the nave were completed in the first decades of the thirteenth century, as was much of the west front; a dedication of the structure dates to 1239. The upper portions of the west front continued to be refined into the fourteenth century. The Wells chapter house is a late-thirteenth-century addition, and the completion of the Lady Chapel and retroquire dates to the fourteenth. Tudor-Craig identified six types of foliate sculpture at Wells.

[239] The early work of the Wells eastern bays transepts is often omitted in historical narratives of Early English Gothic architecture, which tend to begin at Canterbury Cathedral and continue to Saint Hugh's choir at Lincoln of the 1190s. The placement of Wells usually occurs later chronologically to account for its significant west front built in the second quarter of the thirteenth century. The early work at Wells, however, is significant and represents an approach to architectural design and foliate carving that is entirely separate from contemporary work at Canterbury; its significance has perhaps been undervalued.

In the west arcades of the main transept arms and the first bays of the nave belong to a

separate phase of construction; these capitals integrate figurative carving of heads, Biblical

figures, and playful scenes in the midst of the foliage, including a *spinario* figure (fig. 5.53).[240]

In some capitals in the transept, figures and heads emerge from the vertical stems (fig. 5.54,

5.55), while others employ additional projecting foliate buds from the capital's stalk (fig. 5.56).

In these capitals, the foliate forms begin to unravel and lose their bulbous appearance, sometimes

with a single leaf motif in low relief below. These capitals set the stage for the last collection of

foliate capitals that adorn the westernmost piers in the nave in their treatment of foliage. In the

eastern arcade of the main transepts, these bulbous forms are articulated even further, projecting

outward at a more extreme curvature and occasionally incorporating a single-leaf element in the

lower register of the capital (fig. 5.58).

The west pier capitals lose all sense of the bulbous form and, in effect, "explode" the

"crocket" into a conflagration of twisting abstracted leaf forms (fig. 5.59-61). One of the most

striking characteristics of the capitals crowning the compound piers in the nave at Wells is their

lack of coherent symmetry; each group of three bundled shafts has been executed individually,

giving the whole capital an appearance of unruliness and controlled disorder. While some

capitals maintain a hook-like element that could be classified as a crocket, others leaf out

unexpectedly or appear tossed in an imaginary, multidirectional wind. Some of the foliage

projections of these capitals even extend past the boundary of the abacus, which makes the

foliage take on an apparent visual force. Progressing towards the west end of the nave, the foliate

capitals of Wells become ever more elaborate: the foliage curves under, diagonally, in two or

---

[240] Matthew M. Reeve, "The Capital Sculpture of Wells Cathedral: Masons, Patrons and the Margins of Gothic Architecture," *Journal of the British Archaeological Association* 163 (2010): 72–109. Reeve focuses primarily on the figurative sculpture at Wells, framing the foliate capitals in their increasing complexity as an exemplar of social competition among masons.

three layers, splits off, sprouts again. Whereas pattern and variation can be readily identified in continental examples, the lush foliage at Wells is a fantasy of foliage, almost as if it were threatening to take over the pier itself—a foreshadowing of the design of the west front's singularly unique foliate architecture, which appears to expand an idea developed in the construction of Lincoln Cathedral.

Like the eastern bays of the choir at Wells, foliate capitals from St. Hugh's choir at Lincoln incorporate two levels of bulbous foliage with innovative circular abaci. Even with this change in design, sculptors at Lincoln chose this form to decorate the capitals of the choir, including the capitals that crown the crocket piers of the transepts (fig. 5.62, 5.63).[241] The crocket pier is composed of detached Purbeck marble shafts and hexagonal shafts enclosing an inner pier, which is decorated with crockets that appear to travel upwards. The crockets themselves are simply composed: systematic modules of stone that face out towards the diagonals of the pier, resting behind the Purbeck marble shafts (fig. 5.64). The simplified forms of the pier contrast with the foliate forms of the capital, raising questions about the relationship between the crocket capital design in England and its import for the innovation of the crocket pier at Lincoln. In this first iteration, these forms are clearly differentiated. The overall effect, however, is a play of materiality, positive and negative space, as well as architectural and natural form. Not only do these piers openly display this negative space, but they also employ an abstract natural motif creating the effect of a stone pier sprouting upwards from its base. Despite the contrasts, this is a critical point of transfer; the abstract form covers not only the sculpted

---

[241] Paul Binski has pointed to this natural metaphor as an architectural manifestation of Aaron's Rod and the idea of the blossoming church that comes from Isaiah 27:6 and 35:1. Jean Bony thought this form derived from chevron moldings found at Canterbury Cathedral. The sculptors at Lincoln are notably experimental with surface texture in the eastern bays of the nave, St. Hugh's choir, and—without exception—the Angel Choir. See also Frankl, *Gothic Architecture,* 101; Carolyn Marino Malone, *Façade as Spectacle: Ritual and Ideology at Wells Cathedral* (Boston: Brill, 2004), 29.

field of the capital, but the architectural support itself, introducing a new kind of living architecture wrought in stone in England.

The Wells west façade is the centerpiece of the last phase of construction at Wells during the tenure of Bishop Jocelin (fig. 5.65). The great screen-like façade, punctuated by quatrefoils and niches for sculpture, is also alive with crockets. While this is difficult to see in images of the façade that show the building straight-on, the foliate frame springs to life when viewed from an oblique angle, especially in the front buttresses (fig. 5.66, 5.67). In this ensemble, crockets are attached to slender colonnette-like elements, though they are difficult to define since the foliate decoration takes over any sense of vertical support. (fig. 5.68). These crocket colonnettes are, in effect, caged in by smooth, slender shafts that create the architectural frame of the whole ensemble. The crockets cover the entire façade, from top to bottom and right to left. The idea of living architecture forged at Lincoln has been translated from interior to exterior, from occasional support to the entire frame of the western front of the building, engaging with the figurative sculpture in a conversation of vitality and regeneration. The crockets are more elaborate than the simplified forms found at Lincoln; at Wells, the crocket piers are remarkably detailed and diverse, consisting of alternating forms for curling foliage and tri-lobed leaf forms. This decorative sculpture produces the effect of an architecture that appears to grow miraculously out of the earth, framing what was to be an unprecedented sculptural program, complete with quatrefoils carved in high relief. The crocket elements are complemented by flourishes and foliate tympana that fill the niches in the upper portions of the façade, which echo foliate tympana integrated into the nave triforium (fig. 5.69).

The architectural relationship between Wells and Lincoln has long been acknowledged; the two dioceses also had a familial relationship in the early thirteenth century, as Bishop Jocelin was brother to Hugh, Bishop of Lincoln.[242] The crocket piers used in the west façade at Wells

supersede mere architectural quotation; Thomas Lorreys, the architect who took over

construction after the death of Adam Lock in 1229, has transformed this element from a novel,

but rather marginal, feature to an entire architectural framework. Foliate tympana, like their

counterparts in the nave interior, further enhance the vegetal look of the structure. Notably,

iterations of the Wells west front design at Salisbury and Trondheim reproduce the screen-like

architecture, but do not integrate this approach to sculpted foliage; the ensemble at Wells is truly

unique.

The unique integration of upward-growing crockets in the façade at Wells may have been

a unique approach to foliate sculpture that expressed a local identity. Like many sacred sites that

became Christian churches, Wells was located by famous springs, first referenced in a charter of

766.[243] At Wells, however, the connection with the miraculous springs was maintained not only

through the name of the city, but the medieval images that contributed to the formation of the

city and cathedral's identity. The borough of Wells depicted the cathedral, springs, and growing

foliage on its thirteenth-century seal (fig. 5.70). Warwick Rodwell has examined the connection

between the cathedral's early history and the depiction of the springs, which appear beneath three

arches below the depiction of the cathedral, but the decision to include growing foliage—

traditionally understood to be an ash tree—has not been examined in concert with the unique

foliate program of the western façade at Wells. While the presence of sacred springs predating

the construction of churches and cathedrals occurs frequently in medieval Europe (and elsewhere

in the medieval world), the decision to connect the cathedral's identity to these springs through

---

[242] Malone, *Façade as Spectacle*, 29.
[243] Rodwell, *Well Cathedral*, 375.

official imagery and the suggestion of miraculous growth on the borough's legal seal is unusual in the thirteenth century.

With its experiments in bursting crocket capitals as well as the use of the crocket pier as a significant architectural element in the west façade, the work at Wells also appears to foretell trends in sculpted foliage that would emerge later in the thirteenth century, both in England and across the Channel. In France, many crocket capitals from the mid-thirteenth century terminate not in hooks, curls, or buds, but clusters of burgeoning foliage. While this transformation has been explained as an organic process of inward-looking stylistic development that expresses *le génie* of the medieval French sculptor, the dynamic conversation between sculptors and patrons in England and France—as well as the incidental events of creativity and innovation—should perhaps play a larger role in parsing out these narratives. Exuberant foliage that articulates architectural surfaces was an important organizing principle in the design of the chevet at Ely (fig. 5.71) and the Angel Choir at Lincoln (fig. 5.72, 5.73). In both of these examples, bursting crockets fill the spandrels, shafts, and window moldings, producing an effect of glittering vitality.

5.6 The Church as a Living Architecture

Despite its novelty, the appearance of the crocket pier at Lincoln bears some formal resemblance to the hewn cross of the carved Cloisters Cross thought to have originated from Bury St. Edmonds (fig. 5.74). Noted for its singularity as well as its anti-semitic Latin inscriptions, the walrus ivory cross presents a visual manifestation of the living wood of the cross, an image connected with the Tree of Life and the Tree of Knowledge through the inclusion of Adam and Eve at the cross's foot. This type of hewn cross appears in several manuscript illuminations of the twelfth and thirteenth centuries, and the formal resonance between the textured piers and the vertical shaft of the hewn crosses, which were also illustrated in manuscript illuminations, is an attractive invitation to draw a comparison to the crocket piers at Lincoln and Wells.[244] The visual similarity notwithstanding, however, the crockets on the architectural piers at Lincoln and Wells project outward into space, curl over, and create an effect of upward, organic growth, while the jagged, hewn wood of the Cross in these images produces a different visual effect and meaning. Would a "hewn pier" possess the same upward movement and gestures towards foliate embellishment?

Perhaps foliate crosses, such as the manuscript illumination from the Psalter of Robert of Lindsey, or the *crux gemmata,* present a more productive comparison.[245] Paul Binski has highlighted the etymological slippage between the two definitions of *gemma,* "gem" or "bud," to connect these traditions of representation of the Cross in medieval art, and the connection could quickly transfer to crockets, especially those that suggest budding plants. The definition of the term in ancient and medieval texts, however, is made clear by context; it draws from the Greek γέμω, "to be full," indicating a fullness or swelling. The most common usage in Latin texts refers clearly to jewels and precious stones; in Suger's writings, this is also the case. There is no indication that the abbot plays on this usage of *gemma*; his context in describing the precious materials (and omitting details about capital carving or foliate details) is clear; the Latin usage of *gemma* that denotes "bud" is less common in the literature. I am not aware of any evidence that shows that this term was used to describe crocket piers, crocket capitals, or foliate sculpture of any kind. While the resonance between crocket piers, foliate crosses, and *crucis gemmatae* connects these traditions through material richness and artistic vitality, the architectural

---

[244] Jennifer O'Reilly, "The Rough-Hewn Cross in Anglo-Saxon Art." In *Ireland and Insular Art, A.D. 500-1200*, edited by Michael Ryan. Dublin: Royal Irish Academy, 1987. pp. 154, 157.
[245] Crucifixion, Psalter of Robert of Linsey, Society of Antiquaries MS 59, fol. 35v, c1220 (Society of Antiquaries, London). Reproduced in Binski, *Becket's Crown,* p. 215.

manifestation of this idea has a distinct power in defining sacred space. While some art historians

have used the crocket pier as an accessory to explaining these other phenomena, perhaps it is the

crocket pier itself that should be centered in these discussions, especially because it appears in

architecture as early as the 1190s. Crockets might have a resonance with depictions of the living

wood of the Cross, but they are also representations of living stone—and by extension, a living

architecture.

This living architecture could be considered a backdrop for writers like St. Bonaventure,

who work *Lignum vitae* in the latter half of the thirteenth century. This text contained forty-eight

short meditations on the mysteries of Christ, in which the cross of the crucifixion and the Tree of

Life in Paradise are both Christological symbols, recalling the image of Adam and Eve at the

foot of the Cloisters Cross. An image of a tree with the crucified Jesus accompanies many

surviving manuscripts, with branches symbolizing events from his life and fruits symbolizing the

twelve apostles. The earliest of these images appears in a manuscript now in the British Library,

dating to the third quarter of the thirteenth century, though Bonaventura's text was translated into

vernacular and disseminated widely in the fourteenth and fifteenth centuries—and the author

would not achieve sainthood until 200 years after his death. Speaking to the image of the tree,

however, Bonaventura knew the power of natural metaphor in communicating ideas to a wide

audience. In connecting the cross and the Tree of Life in Paradise, Bonaventura recognized that

the tree itself was an image that could be understood by "simple, ordinary people who lack

higher education,"—in other words, the same audience St. Bernard describes as needing to be

moved to religious devotion by images in the more public setting of the cathedral, as described in

the introduction.[246]

---

[246] Qtd. in Marianne Schlosser, "Bonaventure: Life and Works," trans. Angelica Kliem, in *A Companion to Bonaventure*, ed. Jay M. Hammond, J. A. Wayne Hellmann, and Jared Goff, Brill's Companions to the

The connection of these textual and artistic traditions with the experiments of imbuing a sense of liveliness in sacred space, however, should be reconsidered. Crocket architecture was common in certain locales by the end of the twelfth century, increasing in its frequency and inventiveness. Vertical crocket elements appear in France in the thirteenth century; the addition of a large west window and figurative sculpture at Vézelay introduced a wealth of crocket embellishments (fig. 5.75). The late-thirteenth century cloister of Saint-Jean-des-Vignes in Soissons abounds in foliate ornament, including crocket colonnettes that climb the exterior of its piers (fig. 5.76). Crocket piers and crocket friezes articulate the exterior of Burgos Cathedral and we might include, as another manifestation of the foliage-laden frame for a grid of sculpted figures, the retrofacade of Reims Cathedral, constructed between 1265-1285, which employs flat expanses of rich foliage.[247] In all of these examples, sculpted foliage has become a defining element not just of the architectural decoration, but the architectural elements themselves. This architectural sculpture is a mode of living architecture that provided fertile ground in which ideas about the natural world, Christ's resurrection, and sacred space could be contemplated within a more reflexive field.

Christian Tradition, volume 48 (Leiden; Boston: Brill, 2014), 15.

[247] Donna L. Sadler, *Reading the Reverse Façade of Reims Cathedral: Royalty and Ritual in Thirteenth-Century France* (Farnham, Surrey; Burlington, VT: Ashgate Publishing, 2012).

6

Conclusion:

Towards a Cathedral of the Environmental Imagination

Should we pay attention to foliate sculpture in Gothic architecture, these seemingly minute or "marginal" elements of architectural decoration? And if so, how deeply should we seek to understand these forms? What can we learn from them? In this dissertation, I have aimed to defamiliarize elements of Gothic architecture that can easily be taken for granted, whether crockets, foliate friezes, or depictions of nature and seasonality. These studies have focused on examples of foliate sculpture that were chiseled in the late twelfth century and the first decades of the thirteenth because of their ingenuity and import. Examining these case studies deepens the understanding not only of the foliate decoration of this period, but also the long durée of foliate sculpture through the end of the thirteenth century and beyond.

6.1 Exchange, Production, and Meaning

Foliate sculpture was a frontier of artistic and cultural production in the Middle Ages, as well as a site of transformation and exchange. Natural metaphors in architecture are, in fact, powerful vehicles for meaning in sacred space. But it matters where this sculpture is placed and how it interacts with the three-dimensionality of the space; at Amiens, we might even include a fourth dimension, time, as the north portal suggests a cosmic unfolding of the seasons interspersed with ongoing ecological miracles like the procession pictured in the portal's tympanum. It is notable that the vegetal look of specific churches preceded certain foliate trends

174

in other media, literature, and cultural phenomena, as in the work of Albertus Magnus, Bonaventura, and Marian paraliturgical texts. Many innovations at Amiens, such as the delicate rose frieze on the west façade, were also adapted to other contexts later in the thirteenth century, as seen at the Porte Rouge at Notre Dame of Paris and the alternating crocket friezes at the Sainte-Chapelle and Notre-Dame of Reims. In relation to Amiens, the portals at Noyon, which have received little scholarly attention, also retain important information about foliate decoration. In looking at what remains at Noyon, the textile-like diapering pattern, quatrefoils, and foliate friezes are strikingly similar to the ensemble at Amiens, which also employs a foliate diapering pattern (called "la mosaïque") in the three western portals as well as in the portal of St. Christopher.

The discourse on foliate ornament and architectural decoration is evolving, as art historians determine how to move forward from discussions of structure and ornament so engrained in the study of architectural history. A central question of architectural criticism, especially for pre-modern buildings, is how the reception of Vitruvius might have shaped the discourses we inherit. As Vitruvius aligns vegetal ornament with the feminine, and complex aspects of gender have played out in criticism, culture, propriety, and "taste," it is useful to revisit these ideas if only to check our own assumptions about what we see when looking at fragments or entire buildings that survive in the pre-modern material record. Vegetal sculpture and foliate forms were expensive to fabricate and maintain, as Viollet-le-Duc's restoration records show, so the intention to include, replicate—and keep—these elements of the building benefits from a wider discourse that accepts a broader range of interpretive possibilities beyond the field's traditional demarcations.

Thinking broadly about foliate sculpture and meaning, we might return to the words of

Durandus and the dedicatory hymn *Urbs Beata Ierusalem*, in which the living stones of the

church evoke the heavenly Jerusalem and paradise on Earth. Foliate sculpture not only attends to

local needs and lore but connects with broader Christian themes of the world to come for the

elect, an attempt to build a representation of heaven in the earthly realm—but also one that, in all

its variation, experimentation, and desire to represent the natural phenomena, is deeply human.

In this way, foliate sculpture could be seen to represent the future, present, and past: the heavenly

ideal as well as an implied threat of expulsion, both Paradise and Garden. In the context of

medieval sacred architecture, foliate sculpture suggests a lithic generative force in its very

stones. This mode of living architecture serves as a representation of grace as well as a petrified

spiritual authority on the natural world, one that is uniquely ordered according to the priorities of

the community as well as the hands of individual artisans. These aspects of church decoration,

plastic but organic, permanent but delicate, not only expand discussions of change and

transformation in medieval sacred space, but through their reflexive relationship to climate,

liturgy, and production, activate new dimensions in the study of Gothic architecture.

There is much to be gained from examining France and England together as a joint

corpus of architectural and sculptural production. Like the high-ranking clergyman at

Canterbury, Lincoln, and Wells who commissioned these vast building projects, craftsmen also

moved back and forth across the Channel with documented frequency. Though we have little

evidence as to the "nationality" of craftsmen and masons who worked on these projects in the

twelfth and thirteenth centuries, medieval sources like Gervase of Canterbury suggest that this

skilled workforce traveled often, following the influx of funds into rebuilding projects and taking

up residence wherever they could find work. While allegiances to and relationships with the monarchy were important in certain building projects, the Church maintained its own political structure that allowed for intellectual and artistic exchange, even in times of political conflict. This project suggests that while sculptors and masons responded to local trends within networks of local workshops, responding to the site and even the local identity of the church or diocese was a top priority in church decoration. The formal comparisons that exist between Wells and Amiens, for example, might be considered in a broader inquiry that considers the architecture and sculptural production across the Channel more broadly.

Even the most modest of foliate articulations, the ubiquitous crocket, opens questions into the adaptation of Gothic architecture across geographic areas. Crockets are, in fact, a crucial element of Gothic architectural design and communicate the very characteristic of adaptation at the heart of the Gothic architectural enterprise in medieval Europe up until the Renaissance. A number of formal elements make up what we have come to call "Gothic," and the crocket must be counted among them. The focus on modular elements and pattern might be considered in a rhetorical framework that connects repeating aesthetics with other cyclical or repetitive aspects of medieval culture, rather than only emphasizing monotony, engineering, and economics.

6.2 Naturalism, Intellectual History, and Foliate Sculpture

This study has been careful to assign meaning based on medieval texts and medieval ideas about the natural world. The written record is, fortunately, rich with insights, showing us how individual medieval scholars grappled with the texts they could access in their monastic libraries as well as their knowledge of local religious traditions, biblical and apocryphal texts, and experience of nature and seasonality within their local contexts. More research could be applied in this area to the chroniclers who recorded meteorological phenomena and the process of the growing seasons, as their records might provide a more detailed, localized view of engagement with nature and understanding of the ongoing changes and processes of the natural world in the Middle Ages. That said, the literature analyzed within this project points to a concerted effort to understand the generative forces of nature as well as a preoccupation with the practical means of harnessing its potential. Twelfth- and thirteenth-century natural philosophy and discourses on nature reflect a strong desire to understand plant life theoretically, scientifically, and spiritually. This literature also reflects an apparent confidence and optimism that these dynamic forces might be understood through careful study, rhetoric, and systematic thinking, as in Walter of Henley's *Husbandry*. This is notable, given the different orientation towards nature found in several early Christian texts, in which resource exhaustion and the end of times paints discussions of nature—and the relationship between man and the natural works of God—in a rather harsh light.

As evidenced by these examples, not all foliate decoration worth studying attends to the rules of naturalistic form. Many important elements of Gothic foliate sculpture are, in fact, not naturalistic but *super*natural, invoking a divine nature that exists only within the context of sacred space. This project has shown that foliate sculpture need not adhere to naturalistic aesthetics in order to be analyzed and incorporated into important aspects of a building's history and study. Building a vocabulary to speak about these elements, in many cases, is still a necessary first step in regaining interpretive footholds to move the discourse forward. Especially in cases where the literature flattens complicated and dynamic phenomena with long-expired terminology or vague language, it is worth re-investigating foliate forms in order to move towards an interpretive framework that brings these important elements of architectural sculpture into the fold of Western art history. These forms engage in a deft slippage between natural and supernatural, bringing viewers into a lithic version of the natural world that mimics its organic

qualities without recreating it in a literal sense. Foliate sculpture in the cathedral is nature

transformed, nature at the hand of the artisan, and nature under the divine protection—and

control—of the Church.

6.3 Foliate Sculpture and the Dynamics of Change

Focusing on the specific, the frieze of Notre-Dame of Amiens is one of the most

significant works of foliate sculpture in Western Europe. With its changing, trilobed leaf forms

and diverse fruits, the foliate frieze at Amiens was composed to heighten the organic quality of

the architecture while providing a reference point for the cathedral's specific local context as the

rightful home of the relics of Saint Firmin and site of important liturgical events. Just like

molding profiles have reflected subtle changes at Notre-Dame of Amiens in the earlier phases of

construction, destabilizing the degree to which we can be certain that stonework was

prefabricated and mass-produced, the foliate frieze also undergoes a number of changes as it

travels across the length of the cathedral's interior. Some of these changes indicate a certain

amount of experimentation, not unlike the foliate carving in the nave capitals in the southeast

bays, but also provoke questions of why the Amiens sculptors never settled on a faithfully

repeating pattern, when other, earlier exemplars at Soissons, Saint-Remi, and Vézelay opted for

replicating motifs. An interest in the aesthetic of vitality and acceptance for the human variation

of the artisan's hands is petrified in this unique example of foliate sculpture that sits between the

classifications of architectural decoration and three-dimensional installation art. While my

analysis stops short of identifying individual "hands" in the production of the frieze due to the

uncertainties of production and other issues with this methodology, further analysis of this sculpture could provide insight as to approximately how many sculptors might have worked on the frieze at any given time, if done with care.

The frieze is a type of festooning, meant to correspond with hanging textiles below, but it also has major implications for the importance of foliate sculpture in sacred space. Not only does the ever-changing composition of the frieze challenge narratives of serial production at Amiens, but its formal qualities also raise questions about the production of meaning within sacred space and the positioning of the natural world in Christian thought. As highlighted in chapter 3, attention to detail in the portrayal of the natural world was an aesthetic priority at Amiens. The unique treatment of foliage suggests not only the entanglement of nature, man, and God, but also a worldview in which the ordering of the seasons, progression of natural processes, and cultivation of plants were subject to divine protection and religious practice. The decision to install a continuous, undulating foliate frieze half a meter tall across the entire expanse of the cathedral represents a significant investment in the cathedral's foliate program as well as an investment in this worldview in which the Church played a key role as arbiter of nature, *animus*, and seasonal change. At the same time, the form of the frieze resists our comprehension from the ground floor of the cathedral, the vantage point from which it was meant to be viewed, ever increasing its mystery and miraculous nature. The fact that the forms of the frieze have thwarted even the most careful art historians deserves pause; its composition is only realized in this project with the help of a telephoto lens, digital photography, and enough computing power to process image files large and detailed enough to capture it.

This project has also shown how to work with and against existing theories of stylistic change, focusing on the objects *in situ* and the ways in which they interact within the hyperlocal context of a specific sacred space. At Amiens, the revolution of the frieze reaches its end in the choir, where it is more a denouement: in the third and final installment of the frieze, the strong directional emphasis is given over to a repeating pattern, delicately wrought but far less robust than the other two sections in the nave and transept arms. The change in directional emphasis is important and represents a reorientation to this sculpture element and its role in the architectural and liturgical ensemble. In the nave, the frieze appears to travel around the building counterclockwise; in the transept arms, the directionality of the frieze is composed to frame the choir; and in the choir, the plastic qualities of the frieze are diminished by alternating, fluttering leaf forms and blossoming crockets. But the de-emphasis of the frieze's form in the choir does not represent a uniform de-emphasis of foliate decoration altogether in the latter half of the thirteenth century; indeed, it is in this section of the building that some of the most brilliant foliate capitals make their appearance in the chevet dado and delicate triforium. What we see at Amiens is a reordering of aesthetic priorities in this last phase of construction, overseen by master mason Renaud de Cormont. This phase was also impacted by budgetary constraints and political instability, as evidenced by the fire of 1258, its aftermath, and other cost-saving measures that might have eased the burdens of the cathedral fabric, such as the lightweight openwork flyers.

At Wells, on the other hand, foliate intensification is evident from the quire traveling westward, as the supernatural foliate capitals intensify in their multiplicity and complexity, the carvers even abandoning symmetry to explore ever-more complicated manifestations of foliate carving. The climax of the foliate program at Wells, which also includes foliate tympana in the nave and a wide range of foliate capitals and bosses, is the western frontispiece, which is composed of an architectural network of foliate crocket piers that appear to grow upwards. As at Amiens, the attention to foliate detail appears to be linked to a specific local context; here, the eponymous wells located at the site of the sacred space, the iconography of which incorporates growing foliage or an ash tree. Also like Amiens, most of the sculpted foliage at Wells was not created in a naturalistic mode, but an other-worldly aesthetic that supersedes mere abstraction or stylization.

6.4 Climate and Sacred Space

Records of medieval environmental miracles, which exist in large numbers, yield important information about how church authorities entangled religious belief and natural phenomena, especially in the ways local saints were invoked to intervene in the wake of ecological disasters. Putting this qualitative evidence together with paleoscientific data that documents, with increasing certainty, changes in medieval climate is crucial in approaching the question of whether changes in climate in Northern Europe impacted patterns of religious practice and devotion—and, by extension, the way in which clerics, patrons, and artisans chose to build and adorn sacred space. Generally, centering ecological phenomena in historical and art historical discourses presents new questions about behavior, religious belief, and artistic expression in the medieval world. Church building, in its materiality, labor force, aesthetic transformations, and even liturgical use, is very much entangled with climate, weather, and ecological disaster. Broadly speaking, the case studies documented in the dissertation were conceived and manufactured during an era of remarkable confidence and prosperity; they show a preoccupation with the non-animal, non-human world on the part of artisans and designers of iconographical programs. They show a sacred space blooming organically from its very stones; at the same time, they show the threat of desolation and damnation.

Foliate sculpture sits dynamically within the tripartite rhetoric of creation: God as creator, man as artisan, and works of nature created by a generative force, or *animus*. As foliate carving is the work of the artisan, it is the work of man; as this sculpture represents works of nature, it references the generative *animus* behind that natural world, created by God. Foliate carving from the twelfth and thirteenth centuries is not only carefully composed, but also made to appear *cultivated* by a divine hand. Given the proliferation of foliate carving in certain buildings, the carved natural forms of the cathedral also mimic the careful tending of fields and forests documented in medieval texts. We see illustrations of willful control in many examples: in the perfection of the manicured vine, and the ordered symmetry of French crocket capitals. It is tempting to read Lynn White Jr.'s thesis about Western culture's preoccupation with controlling and exploiting nature into these particular examples, in the way they occupied a discrete field, retain symmetry, or perfectly geometrical formal qualities.[248] And indeed, the desire to understand the workings of nature, the passage of seasons, the cultivation of specific plants, and successful agricultural techniques is reflected in Western medieval literature and text circulation.

But the visual record of foliate sculpture holds a more nuanced view: in sacred space, nature was held in balance between human intervention and divine control. The western façade at Amiens provides ample examples of desiccated plants and evil trees as well as sheltering, bountiful foliage and trees divinely favored. The importance of the north portal and iconography of the Invention of Saint Firmin in relegating this foliate program has been, in my view, underestimated. The warming miracle of Saint Firmin visually interrupts the winter months in the quatrefoil calendar—an iconographical positioning unlike any other example of portal sculpture incorporating the labors of the months and zodiac. In a visual relationship with Christ

---

[248] Lynn White, Jr., "The Historical Roots of Our Ecologic Crisis," *Science* 155, no. 3767 (1967): 1203–7.

in the central portal and the Virgin Mary in the south portal, Saint Firmin is presented as the local Christ figure powerful enough to intervene in the natural order. As such, the cathedral lays claim to the power of Saint Firmin's relics, which, given the events of the twelfth century and relic controversy at Saint-Denis, reinforces the importance of this miracle and the foliate program in the cathedral's architecture.

Whether medieval makers made the connection between *animus* and their depictions of the natural world, it is hard to say. As explored in chapter 2, Albertus Magnus was among the first to connect the study of individual plant species with a philosophical worldview, though he did not complete this work until the end of his life. Furthermore, *De vegetalibus* was not circulated widely until after his death; the proliferation of translations appeared in scriptoria in the fourteenth and fifteenth century, while the examples of foliate sculpture in this study are clustered around the decades before and after 1200. This project has shown that foliate activity in architectural carving predated many of the more intense and novel engagements with plant life in Western medieval literature, leading me to posit that the vegetal carving and intellectual activity were separate, but related, threads of a complicated history of thought. If one informed the other, it seems more likely that foliate carving appearing in the latter half of the twelfth century and early decades of the thirteenth pushed Western medieval thinkers to consider plant life as a subject of study both scientific and divine in the years towards 1300 and beyond.

Western medieval art history benefits from the study of nature and climate in the twelfth and thirteenth centuries. This was an era of remarkable confidence that nature could be controlled through human intervention—not just in agriculture, but devotional practice as well. We are left to question how the experience of the actual medieval climate impacted culture and artistic production at large. In the economic expansion of the twelfth and thirteenth centuries, the advancements in technology were aided by favorable growing seasons that lasted for decades in northern Europe during the Medieval Warm Period, which coincided with what historians have considered to be demographic expansion and economic growth. These were also more favorable conditions for major, resource-intensive building projects that include expansive public works, such as the city walls of Paris as well as the concurrent construction of major cathedrals in northern France and England, Notre-Dame of Amiens and Wells among them.

The impact of these conditions has not been centered in much of the discourse or in discussions of medieval intellectual activity in the twelfth and thirteenth centuries. Given our current experiences of climate instability, it is notable that the "age of the great cathedrals" fades in the decades following 1300, a period that coincides not only with the gradual end of the Medieval Warm Period and the beginning of unfavorable growing seasons that contributed to the Great Famine of 1315-17, a notable ecological and economic disaster that many medieval historians consider to be of greater consequence than the Black Death in reshaping medieval society. Broadly speaking, the relative dearth of large cathedral-size building projects and proliferation of smaller church expansion projects such as porches, Lady Chapel extensions, chapter houses, and lateral chapels in fourteenth-century France and England is often explained away by economic and political conflicts. The connection between the instability leading up to the Hundred Years' War and the climate-related famine that preceded it, however, is worth taking seriously, especially considering how economic conditions might have informed the resource availability and the general will to plan, fund, and undertake large building projects.

Perhaps this is what the foliate chevet at Amiens portends, a reflective turn towards the smaller, delicate foliate details in the small capitals as opposed to robust, ambitious projects like the carved frieze. Considering this history and this moment in time, it is difficult to ignore the direct impact of climate on cultural and artistic production. The global Covid-19 pandemic brought the world to a halt in March 2020. In the week that I wrote this conclusion in New York City, dozens of people drowned in climate-change-related flooding unlike anything we have seen in the tri-state area. If we look to the science, an era of climate disasters awaits us. How cultures before us dealt with natural disaster and used art as a way of mitigating the natural world demands our attention, especially at this moment. I hope we will continue to look at these examples and remain open to what they have to teach us.